LONDON MATHEMATICAL SOCIETY S

D0147571

Managing Editor: Professor D. Benson,
Department of Mathematics, University of Aberdeen, UK

London Mathematical Society Student Texts 87

Riemann Surfaces and Algebraic Curves

A First Course in Hurwitz Theory

RENZO CAVALIERI
Colorado State University

ERIC MILES
Colorado Mesa University

CAMBRIDGE
UNIVERSITY PRESS

CAMBRIDGE
UNIVERSITY PRESS

One Liberty Plaza, 20th Floor, New York NY 10006, USA

Cambridge University Press is part of the University of Cambridge.

It furthers the University's mission by disseminating knowledge in the pursuit of education, learning, and research at the highest international levels of excellence.

www.cambridge.org
Information on this title: www.cambridge.org/9781107149243

First published 2016

Printed in the United States of America by Sheridan Books, Inc.

A catalogue record for this publication is available from the British Library

Library of Congress Cataloging-in-Publication data
Names: Cavalieri, Renzo, 1976– | Miles, Eric (Eric W.)
Title: Riemann surfaces and algebraic curves : a first course in
Hurwitz theory / Renzo Cavalieri, Colorado State University,
Eric Miles, Colorado Mesa University.
Description: New York NY : Cambridge University Press, [2017] |
Series: London Mathematical Society student texts ; 87
Identifiers: LCCN 2016025911 | ISBN 9781107149243
Subjects: LCSH: Riemann surfaces | Curves, Algebraic. | Geometry, Algebraic.
Classification: LCC QA333 .C38 2017 | DDC 515/.93–dc23
LC record available at https://lccn.loc.gov/2016025911

ISBN 978-1-107-14924-3 Hardback
ISBN 978-1-316-60352-9 Paperback

Title page: *Topology Tree*, by Kris Barz, 2015. http://krisbarz.squarespace.com

Contents

Introduction

Hurwitz theory is a beautiful algebro-geometric theory that studies maps of Riemann Surfaces. Despite being (relatively) unsophisticated, it is typically unapproachable at the undergraduate level because it ties together several branches of mathematics that are commonly treated separately. This book intends to present Hurwitz theory to an undergraduate audience, paying special attention to the connections between algebra, geometry and complex analysis that it brings about. We illustrate this point by giving an overview of the material in the book.

Hurwitz theory is the enumerative study of analytic functions between Riemann Surfaces – complex compact manifolds of dimension one. A Hurwitz number counts the number of such functions when the appropriate set of discrete invariants is fixed. This has its origin in the 1800s in the work of Riemann, who first had the insight that multi-valued inverses of complex analytic functions can be naturally seen as functions defined on a domain which is locally, but not globally, identifiable with the complex plane: i.e. a Riemann Surface.

Studying analytic functions defined on Riemann Surfaces leads to the geometry of oriented topological surfaces, which Riemann Surfaces are. The local behavior of functions reveals a high degree of structure: analytic functions are ramified coverings; that is, coverings except at a discrete set of points where a phenomenon called *ramification* occurs.

Ramified coverings naturally give rise to monodromy representations, which are homomorphisms from the fundamental group of the punctured target surface to a symmetric group. The ramification at the preimages of a point b in the base is captured by the cycle type of the permutation associated with a small loop winding around the point b.

The count of all such representations can be identified with a coefficient of a specific product of vectors in the class algebra of the symmetric group: with a vector space which has a basis indexed by conjugacy classes. Elements of this basis are given by formal sums of all permutations in the same conjugacy class. A commutative multiplication is then defined by extending the group operation of the symmetric group by bilinearity.

The class algebra is known to be semisimple: it admits a basis with respect to which multiplication is idempotent. Computing the product above in the semisimple basis yields closed formulas for Hurwitz numbers in terms of characters of the symmetric group.

To summarize, the count of analytic functions was translated to a geometric count of topological covers, then to an algebraic count of group homomorphisms, and finally reduced to a representation theoretic computation.

In a different direction, Riemann Surfaces can be degenerated to nodal surfaces by shrinking loops. These nodal surfaces look like "smaller" Riemann Surfaces glued at points, and so degeneration creates infinite families of recursive relations among Hurwitz numbers. We conclude the book by showing that when Hurwitz numbers are encoded as coefficients of a formal power series (a generating function called the *Hurwitz potential*), some of these recursions translate into partial differential equations that are solved by the Hurwitz potential.

Whether this summary makes perfect sense or no sense at all depends on the background of the reader. In any case, we hope that at least two things are apparent: first, that keywords from several different undergraduate courses have been used; and second, that no exceptionally sophisticated term appeared.

This book arises from two experimental undergraduate courses that the first author taught at Colorado State University in 2014 and 2015. The courses were offered as a follow-up to classes in topology and differential geometry; a main goal was to depart from the structure of a traditional course and offer the students a mode of approaching the study of mathematics closer to that of a researcher facing a new problem.

At a school like Colorado State University, most advanced math majors have typically taken semester-long courses in some of the areas mentioned in the above synopsis, and typically have not taken all those courses. There is some analogy with the situation that mathematical researchers are in when they tackle an open problem. First of all, translation and reformulation of a problem is often a very important tool in mathematical research. Problems that are too difficult when studied in a certain way may become approachable when the point of view is changed. When mathematical researchers translate a question in order to find ways to solve it, they are often taken into mathematical areas out of their comfort zone. And they don't have the opportunity to take a semester-long course, or to read a whole book on each topic that they use, but must be able efficiently to develop a working understanding of the aspects needed for their problem.

This analogy informed the way we structured the narration of our story. We have background chapters that introduce complex analysis, manifolds, the fundamental group, representation theory of the symmetric group and generating functions in a skeletal way, touching only on content that we considered essential to our scope. Such background is not collected all together at the beginning, but is introduced at the moment when it is needed in the story, which we believe develops the exposition in a more organic way.

We made the choice of having exercises interspersed in the narration of the book, serving as an integral part of the exposition, rather than collecting exercises at the end of each section. The exercises are designed to develop familiarity with the concepts introduced, which is necessary before using the concepts in new ways. Exercises also appear in proofs, partly to avoid the excessive proliferation of parts of proofs that consist mostly in bookkeeping, but also to encourage the reader to be actively involved and test his/her understanding.

This book can and should be used differently by different readers, but we hope that, whether you are an instructor preparing a course, a student reading this independently, or something in between, you find this book a helpful guide through the first steps in this fascinating topic.

Although the main body of the text covers a lot of ground, this is really only the beginning of the story in Hurwitz theory. By nature, Hurwitz theory is interdisciplinary and is part of the basic toolkit in many areas of mathematics. In the appendices we offer a glimpse of what is beyond through a small number of essays by guest writers: active researchers in various areas of mathematics who use Hurwitz theory in their work. The scope of the appendices is to pique the reader's interest; to leave them a bit dazed and confused, and with the desire to continue learning – which is the constant state of mind of any mathematician.

Acknowledgments

Thanks first and foremost to the students at Colorado State University who served as "guinea pigs" for this experiment: Dean Bisogno, Christie Burris, Tucker Manton, Will Piers, Rachel Popp, John Ramsey, David Reynolds, Kyle Rose, Pat Severa, Matthew Smith, Gavin Stewart and Trent Woodbury. Their enthusiasm for learning and hard work made these classes not only fun to teach, but also a learning experience for the instructor. A special mention must be made of Amanda Rose and Nate Zbacnik, who served as "class czars" and kept it all together and organized.

Thanks to Hannah Markwig for testing the book in her class at University of the Saarlands and providing useful feedback.

Kudos to David Allen for producing all the beautiful figures.

1

From Complex Analysis to Riemann Surfaces

This chapter makes a quick and targeted incursion into the world of complex analysis, with the goal of presenting the ideas that historically led to the development of the notion of a Riemann Surface.

Differentiability for a function of one complex variable imposes considerably more structure than the analogous notion for functions of a real variable. Somewhat strangely, many of the remarkable properties of complex differentiable functions are natural consequences of a construction that somehow "leaves" the complex world: complex functions can be integrated along real paths, and the value of such integrals doesn't change if the path is continuously perturbed while fixing the endpoints.

This phenomenon leads to Cauchy's formula, which expresses a complex differentiable function as a path integral of yet another complex function. While this may seem a slightly bizarre thing to do, Cauchy's formula has a number of remarkable consequences. In particular it gives a differentiable expression for the local inverse to a complex differentiable function at a point where the derivative does not vanish.

When a differentiable f function is not injective, obviously there does not exist a global inverse function. However, at any point where f' doesn't vanish, one has multiple local inverse functions (or historically one said that the inverse of f is a multivalued function) and, further, there is a natural way to view all these local inverses as part of a global function defined on a space which, around any point, "looks like" the complex numbers but globally may be different from $\mathbb{C}$. Such spaces are examples of Riemann Surfaces.

In this chapter, which is meant to illustrate how the concept of Riemann Surfaces was developed, we limit ourselves to exploring this picture for the power functions $w = z^k$ and their inverses (the k-th root "functions"). While this may seem unimpressive, Lemma 1.4.4 shows that the power functions, up to appropriate changes of variables, describe the behavior of any

1

holomorphic function around a critical point – a point where the derivative vanishes.

Complex analysis is a beautiful and rich subject, and there is no way that we can do it justice in a handful of pages. We have made the choice of taking a path through the subject that gives a working understanding of a small selection of ideas that are important for the development of our story. We refer the reader interested in further reading to any textbook in complex analysis; for example, Conway (1978).

1.1 Differentiability

The definition of differentiability for functions of one complex variable is in complete analogy with the real variable case.

Definition 1.1.1. A function $f : \mathbb{C} \to \mathbb{C}$ is **differentiable** or **holomorphic** at a point $z_0 \in \mathbb{C}$ if and only if the following limit exists:

$$\lim_{|h| \to 0} \frac{f(z_0 + h) - f(z_0)}{h} = L \in \mathbb{C}. \qquad (1.1)$$

The complex number L is the value of the derivative of f at z_0, denoted by $f'(z_0)$. A function f is differentiable on a domain $U \subseteq \mathbb{C}$ if it is differentiable at every point $u \in U$.

Because the complex numbers are two-dimensional over the real numbers, there are many ways for a complex variable h to approach 0. Hence, the existence of the above limit imposes greater structure on functions of a complex variable. For instance, we have the following properties, which we prove in Section 1.3:

1. If $f : \mathbb{C} \to \mathbb{C}$ is differentiable in a neighborhood U of z_0, then it is infinitely differentiable in U.
2. If $f : \mathbb{C} \to \mathbb{C}$ is differentiable at z_0, then it is **analytic**, meaning that the Taylor expansion of f at z_0 always converges to f in a neighborhood of z_0.

The statements above are not true for functions of a real variable, as illustrated in the following exercise.

Exercise 1.1.1.

1. Construct a function $f : \mathbb{R} \to \mathbb{R}$ such that $f'(x)$ exists and is a continuous but not differentiable function.

2. Consider the function

$$g(x) = \begin{cases} 0 & \text{if } x \leq 0 \\ e^{-1/x^2} & \text{if } x > 0. \end{cases}$$

Show that g is infinitely differentiable at 0 and that all derivatives vanish: $g^{(n)}(0) = 0$. This implies that the Taylor expansion of g centered at 0 is identically 0 (and so the Taylor series does not converge to g in a neighborhood of 0).

Writing complex numbers in Cartesian coordinates $z = x + iy$ for $x, y \in \mathbb{R}$, we may write a function $f(z)$ as $f(x, y) : \mathbb{R}^2 \to \mathbb{C}$. Identifying also the codomain of f with $\mathbb{R}^2$, one has $f(x, y) = u(x, y) + iv(x, y)$ for functions $u, v : \mathbb{R}^2 \to \mathbb{R}$. These are the real and imaginary parts, respectively, of f.

Exercise 1.1.2. Find the functions $u(x, y)$ and $v(x, y)$ associated with $f(z) = z^2$. Compute the derivative $f'(z)$ and find its real and imaginary parts.

Theorem 1.1.2 (Cauchy–Riemann Equations). *Let $f : \mathbb{C} \to \mathbb{C}$ be a holomorphic function on an open subset $U \subset \mathbb{C}$. Considering $f = u + iv$ as a real differentiable function on $\mathbb{R}^2$, then the following identities of partial derivatives hold on U:*

$$u_x = v_y, \qquad v_x = -u_y. \tag{1.2}$$

Proof One may restrict the difference quotient (1.1) to real paths approaching z_0. If f is differentiable at $z_0 = x_0 + iy_0$, the limit is $f'(z_0)$ independently of the choice of path. We consider a vertical and horizontal path approaching z_0. Letting h approach zero along a vertical path gives (note that here $t \in \mathbb{R}$):

$$f'(z_0) = \lim_{t \to 0} \frac{u(x_0, y_0 + t) + iv(x_0, y_0 + t) - (u(x_0, y_0) + iv(x_0, y_0))}{it}$$

$$= v_y(x_0, y_0) - iu_y(x_0, y_0). \tag{1.3}$$

Similarly, letting h approach zero horizontally yields

$$f'(z_0) = \lim_{t \to 0} \frac{u(x_0 + t, y_0) + iv(x_0 + t, y_0) - (u(x_0, y_0) + iv(x_0, y_0))}{t}$$

$$= u_x(x_0, y_0) + iv_x(x_0, y_0). \tag{1.4}$$

Equating the real and imaginary parts from the two computations gives the result. $\square$

The following corollary of the Cauchy–Riemann equations will be extremely important in our story.

Corollary 1.1.3. *Let f be a non-constant holomorphic function. As a function from the real plane to itself, f is orientation-preserving.*

Sketch of proof Intuitively, an orientation of the plane amounts to specifying the notions of "clockwise" and "counterclockwise". Formally, one defines an orientation as an equivalence class of bases of $\mathbb{R}^2$, where two bases are equivalent if the determinant of the change of basis matrix is positive. As a consequence, a function $f = u + iv : \mathbb{R}^2 \to \mathbb{R}^2$ is then said to be orientation-preserving if, on an open dense set of the domain of definition, the determinant of the Jacobian matrix

$$J(f) = \begin{bmatrix} u_x & u_y \\ v_x & v_y \end{bmatrix}$$

is positive. If f is complex differentiable, then the Cauchy–Riemann equations imply

$$\det J(f) = u_x v_y - v_x u_y = u_x^2 + v_x^2 \geq 0,$$

and since f is not constant the inequality is strict on a dense open set.   □

We conclude this section by stating without proof an important property of holomorphic functions. A proof is found, for example, in Conway (1978, Chapter IV, §7).

Theorem 1.1.4 (Open Mapping Theorem). *A non-constant holomorphic function f is* **open**: *if U is an open subset of $\mathbb{C}$, then so is $f(U)$.*

1.2 Integration

Complex analytic functions can be integrated along paths in $\mathbb{C}$ (see Figure 1.1). For a path $\gamma : [a, b] \to \mathbb{C}$ define

$$\int_\gamma f(z)dz = \int_a^b f(\gamma(t))\gamma'(t)dt. \tag{1.5}$$

Example 1.2.1. We compute $\int_\gamma \frac{1}{z}dz$ for γ, a circle of radius r centered at zero. We have $\gamma(t) = re^{2\pi it}$ and $\gamma'(t) = 2\pi i re^{2\pi it}$ for $t \in [0, 1]$. Then

$$\int_\gamma \frac{1}{z}dz = \int_0^1 \frac{1}{re^{2\pi it}} 2\pi i re^{2\pi it} dt = \int_0^1 2\pi i dt = [2\pi it]_{t=0}^{t=1} = 2\pi i.$$

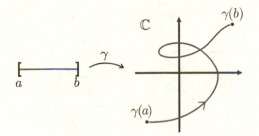

Figure 1.1 Picture of a path in $\mathbb{C}$

Exercise 1.2.1. Show that, for any integer $n \neq -1$, the integral $\int_\gamma z^n dz = 0$, where γ is a circle of radius r centered at zero.

A remarkable property of complex path integrals is that they are invariant under continuous deformations of the path. Intuitively, this means that we can wiggle the circle γ in Example 1.2.1 to be any simple closed curve around the point 0 and still obtain the same result.

We will formalize and discuss extensively the notion of continuous deformation of paths (technically, **homotopy**) in Chapter 5. For now, if γ, η : $[a, b] \to U \subseteq \mathbb{C}$ are paths with the same endpoints (i.e. $\gamma(a) = \eta(a) = z_a$ and $\gamma(b) = \eta(b) = z_b$), a continuous deformation of γ into η is a continuous function $H : [a, b] \times [0, 1] \to U \subseteq \mathbb{C}$ satisfying $H(s, 0) = \gamma(s)$ and $H(s, 1) = \eta(s)$. We also ask that for every t, $H(a, t) = z_a$ and $H(b, t) = z_b$.

The idea is that at time $t = 0$ one has the path $\gamma(s)$, and as time t flows from 0 to 1, the path $\gamma(s)$ continuously morphs into the path $\eta(s)$ while both endpoints stay fixed (see Figure 1.2).

Theorem 1.2.2. *Suppose that $\gamma, \eta : [a, b] \to U \subseteq \mathbb{C}$ are related by a continuous[1] deformation of paths. Then for any holomorphic function f on U, we have*

$$\int_\gamma f(z)dz = \int_\eta f(z)dz.$$

Proof For any $t \in [0, 1]$ we integrate the function $f(z)$ along the path $H(s, t)$, obtaining the function $Int(t) = \int_{H(s,t)} f(z)dz$. Consider the derivative of $Int(t)$ with respect to t:

---

[1] We note that our proof requires the stronger condition that $H(s, t)$ has partial derivatives.

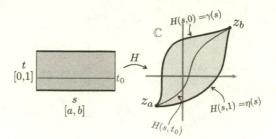

Figure 1.2 Schematic picture showing a homotopy of maps

$$
\frac{d}{dt} Int(t) = \frac{d}{dt} \int_a^b f(H(s,t)) \frac{\partial H}{\partial s}(s,t) ds
$$

$$
= \int_a^b \left(f'(H(s,t)) \frac{\partial H}{\partial t}(s,t) \frac{\partial H}{\partial s}(s,t) + f(H(s,t)) \frac{\partial^2 H}{\partial s \partial t}(s,t) \right) ds
$$

$$
= \int_a^b \frac{d}{ds} \left[f(H(s,t)) \frac{\partial H}{\partial t} \right] ds = f(H(s,t)) \frac{\partial H}{\partial t} \Big|_{s=a}^{s=b} = 0,
$$

since $H(a,t)$ and $H(b,t)$ are constant functions of t. Having derivative identically equal to 0, $Int(t)$ is a constant function and

$$
\int_\gamma f(z) dz = Int(0) = Int(1) = \int_\eta f(z) dz.
$$

$\square$

Corollary 1.2.3. *Let U be a simply connected region of $\mathbb{C}$ and f a holomorphic function on U. For any closed path γ whose image is inside U, $\oint_\gamma f(z) dz = 0$.*

Sketch of proof Let us recall that a path is said to be closed if its endpoints coincide. (The little circle on the integral sign is not strictly necessary, but it is a visual aid to emphasize that the integration is along a closed path.) The definition of being simply connected is essentially that any closed path may be continuously deformed to a constant path. The result now follows from Theorem 1.2.2 since integrating any function along a constant path yields 0 as a result. $\square$

Exercise 1.2.2. Let U be an open set in $\mathbb{C}$ and f a holomorphic function on $U \smallsetminus z_0$. For $j = 1, 2$, let γ_j be a path parameterizing a circle centered at z_0 of radius r_j, oriented counterclockwise and completely contained in U. Show that:

$$
\oint_{\gamma_1} f(z) dz = \oint_{\gamma_2} f(z) dz.
$$

In other words, the value of the path integral is independent of the radius of the circle.

1.3 Cauchy's Integral Formula and Consequences

From the invariance of path integrals under deformation of paths, one obtains a formula for a holomorphic function as a path integral.

Theorem 1.3.1 (Cauchy's Integral Formula). *Let γ be a small loop around $z \in \mathbb{C}$ and $f(w)$ a holomorphic function in a neighborhood U of γ. Then*

$$f(z) = \frac{1}{2\pi i} \oint_\gamma \frac{f(w)}{w - z} dw. \tag{1.6}$$

A formal proof of formula (1.6) may be found in any complex analysis book; for example, Conway (1978, Chapter IV, §5). Let us briefly consider how we should think of this formula and why we should believe it. For any $z \in U$ we intend to describe the value of $f(z)$ as a path integral: at this point we consider z a fixed complex number and the variable of integration is denoted by w. From Theorem 1.2.2 we may assume that γ bounds a small disk around z, and from Exercise 1.2.2 we may let the radius of the disk shrink to 0 without altering the result of integration. Then the function $f(w)$ restricted to γ tends to the (constant) complex number $f(z)$, whereas the path integral of $1/(w - z)$ is $2\pi i$, as seen in Example 1.2.1.

Remark 1.3.2. Cauchy's integral formula may seem baffling at first: if the goal is to understand the function f, why would one make any progress by replacing it with an integral function, which is a more complicated object, and furthermore an integral that involves f itself as a part of the integrand? The answer is that (1.6) is not used to compute values of f, but to deduce properties of f as a function by exploiting the nice formal properties of integrals. We now illustrate this idea by showing some remarkable consequences of Cauchy's formula.

The first remarkable consequence of Cauchy's integral formula is that any holomorphic function can be expressed, in a neighborhood of any point z_0, as a power series centered at z_0. In the string of equations that follows, we assume at every step that we restrict the domain to an appropriate neighborhood of z_0 as needed:

$$f(z) = \frac{1}{2\pi i} \oint_\gamma \frac{f(w)}{w - z_0 + z_0 - z} dw = \frac{1}{2\pi i} \oint_\gamma \frac{f(w)}{w - z_0} \cdot \frac{1}{1 - \frac{z - z_0}{w - z_0}} dw$$

$$= \frac{1}{2\pi i} \oint_\gamma \frac{f(w)}{w - z_0} \left(\sum_{n=0}^\infty \frac{(z - z_0)^n}{(w - z_0)^n} \right) dw$$

$$= \sum_{n=0}^\infty \left(\frac{1}{2\pi i} \oint_\gamma \frac{f(w)\, dw}{(w - z_0)^{n+1}} \right) (z - z_0)^n. \tag{1.7}$$

Formula (1.7) implies that a holomorphic function is analytic: it is infinitely differentiable and the Taylor expansion about any point z_0 converges to the function in a neighborhood of z_0. Finally, it provides integral formulas for all derivatives of f:

$$f^{(n)}(z) = \frac{n!}{2\pi i} \oint_\gamma \frac{f(w)\, dw}{(w - z)^{n+1}}.$$

Definition 1.3.3. Given n, a positive integer, a complex function f has a **pole of order** n at the point $z_0 \in \mathbb{C}$ if $(z - z_0)^n f(z)$ is holomorphic at z_0 but $(z - z_0)^{n-1} f(z)$ isn't.

Exercise 1.3.1. Show that if f has a pole of order n at z_0, then it admits a **Laurent expansion** at z_0; i.e. in a neighborhood of z_0,

$$f(z) = \sum_{k=-n}^\infty a_k (z - z_0)^k,$$

with $a_{-n} \neq 0$.

Definition 1.3.4. Let f have a pole of order n at the point z_0. Then the **residue** of f at z_0 is the $k = -1$ coefficient in the Laurent expansion of f at z_0.

Exercise 1.3.2. Show that if f has a pole of order 1 at z_0, then the residue of f at z_0 can be computed as the following limit:

$$Res_{z=z_0} f(z) = \lim_{z \to z_0} (z - z_0) f(z). \tag{1.8}$$

Exercise 1.3.3 (Residue theorem). Let $\gamma : [a, b] \to U \subseteq \mathbb{C}$ be a simple closed path, bounding a region denoted W, containing the points $z_1, \ldots, z_m$ (see Figure 1.3). Assume f is holomorphic on $U \setminus \{z_1, \ldots, z_m\}$ and has polar singularities at the points z_j. Show that:

$$\oint_\gamma f(z)dz = 2\pi i \sum_{j=1}^m Res_{z=z_j} f(z). \tag{1.9}$$

1.4 Inverse Functions

An important result for us is the Inverse Function Theorem, which says that a holomorphic function admits a local inverse at any point where the derivative is not zero.

Theorem 1.4.1 (Inverse Function Theorem). *Let $f : U \to \mathbb{C}$ be a holomorphic function and $z_0 \in U$ such that $f'(z_0) \neq 0$. Then there exists a neighborhood V of $f(z_0)$ and a holomorphic function $g : V \to \mathbb{C}$ such that $z_0 \in g(V)$ and for every $z \in g(V)$, $g \circ f(z) = z$.*

Proof Since $f'(z_0) \neq 0$ it is possible to restrict the domain of f to an open neighborhood U' of z_0 in such a way that f is injective and f' is never zero on U' [2]. Since non-constant holomorphic functions are open functions, there exists a small ball B_δ centered at $f(z_0)$ with $B_\delta \subseteq f(U')$. Let γ be a path parameterizing the boundary of B_δ. If we let $V = B_\delta$ and restrict f to $f^{-1}(V)$, we have a bijective function that admits a set theoretic inverse. We show that such a function is holomorphic by providing an integral formula for it. For any $w \in V$, define:

$$g(w) = \frac{1}{2\pi i} \oint_\gamma \frac{\zeta f'(\zeta)}{f(\zeta) - w} d\zeta.$$

Let $z \in g(V)$ be such that $w = f(z)$. The integrand of $g(w)$ has a unique pole of order 1 at $\zeta = z$. Applying the Residue theorem:

$$g(w) = Res_{\zeta=z}\left(\frac{\zeta f'(\zeta))}{f(\zeta) - w} \right) = \lim_{\zeta \to z}(\zeta - z)\frac{\zeta f'(\zeta)}{f(\zeta) - w}.$$

Since $f(z) = w$, $\lim_{\zeta \to z} \frac{\zeta - z}{f(\zeta) - w} = \frac{1}{f'(z)}$, giving $g(w) = z$. $\qquad\square$

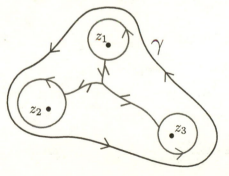

Figure 1.3 Idea behind proof of the Residue theorem

[2] For visual intuition, consider the analogous statement for a real-valued function.

Remark 1.4.2. The Inverse Function Theorem also holds for functions of many variables. A holomorphic function $F : \mathbb{C}^n \to \mathbb{C}^n$ is locally invertible at a point $\mathbf{z_0}$ (with a holomorphic local inverse around $F(\mathbf{z_0})$) if and only if $\det J(F)_{|\mathbf{z_0}} \neq 0$.

1.4.1 k-th Roots

Let $k \geq 1$ be an integer and consider the function $f : \mathbb{C} \to \mathbb{C}$ defined by $w = f(z) = z^k$.

Exercise 1.4.1. Show that, for any $w_0 \neq 0$, the inverse image $f^{-1}(w_0)$ has precisely k elements. The only inverse image of 0 via f is 0.

The derivative $f'(z) = kz^{k-1}$ only vanishes at $z = 0$; by the Inverse Function Theorem, f is locally invertible at every $z \neq 0$. For any $w_0 \neq 0$ and z_0 such that $z_0^k = w_0$ there is a holomorphic function $f_{z_0}^{-1}$ defined in a neighborhood U of w_0 such that $f_{z_0}^{-1}(w_0) = z_0$ and $f \circ f_{z_0}^{-1}(w) = w$ for all $w \in U$. Such a function is called a **branch** of the k-th root function $z = w^{1/k}$ near w_0.

A natural question is: how much can the domain of definition of a given branch of the k-th root be extended? Since $f_{z_0}^{-1}(w)$ provides a choice of a distinguished root for every point of U, one could imagine picking a point close to the boundary of U and repeating the procedure to "enlarge" the domain of definition of $f_{z_0}^{-1}(w)$, perhaps eventually managing to "fill" all of $\mathbb{C} \setminus 0$ (see Figure 1.4). That this is not possible is illustrated by the following exercise.

Exercise 1.4.2. Let $z_0 = 1 \in \mathbb{C}$ identify a branch of $z = w^{\frac{1}{k}}$ near $w_0 = 1$. Consider the path $\gamma : [0, 1) \to \mathbb{C}$ given by $\gamma(t) = e^{2\pi i t}$. What is

$$\lim_{t \to 1} f_{z_0}^{-1} \circ \gamma(t)?$$

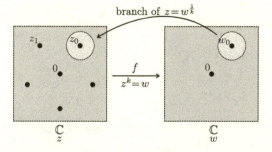

Figure 1.4 Domain and range of $w = z^k$ with branch of k-th root chosen

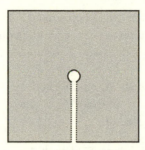

Figure 1.5 A subset of $\mathbb{C}$ where it is impossible to "walk around the origin"

Exercise 1.4.2 shows that a branch of the k-th root may not be extended even continuously to all of $\mathbb{C} \smallsetminus 0$: the issue is that if one follows a branch of the k-th root along a closed path that winds around the origin, one eventually comes back to a different branch of the k-th root. Hence a branch of the k-th root $f_{z_0}^{-1}$ may be extended (as a holomorphic function) to any domain $U \subset (\mathbb{C} \smallsetminus 0)$ such that it is not possible to walk around the origin in U. Typical examples of maximal domains of definitions consist of the complex plane minus a real half-line stemming from the origin (see Figure 1.5).

Another important idea that Exercise 1.4.2 illustrates is that branches of the k-th root should not be thought of as separate objects. One can move continuously from one to another by "talking a walk around the origin", and really there is no particular notion of being on one branch or the next at any given point.

Historically the k-th root was considered a **multi-valued function**, i.e. having k distinct values of z for any given $w \neq 0$. Riemann introduced a shift in point of view by insisting that we should focus on the graph of the k-th power function:

$$\Gamma_k = \{(z, w) \in \mathbb{C}^2 | w = z^k \}.$$

For any point $\mathbf{x} \in \Gamma_k \smallsetminus (0, 0)$, w can be used as a coordinate for Γ_k on a neighborhood of $\mathbf{x}$. Composing with the first projection one obtains a local branch of the k-th root. The space $\Gamma_k \smallsetminus (0, 0)$, which captures all possible branches of the k-th root without making any choice of domain restriction, is called the **Riemann Surface of the k-th root**.

Example 1.4.3. Let us consider the punctured graph $\Gamma_2 \smallsetminus (0, 0)$ of the holomorphic function $w = z^2$, which is the Riemann Surface of the inverse function $z = w^{\frac{1}{2}}$. As a topological space, Γ_2 is homeomorphic to $\mathbb{C} \smallsetminus 0$ (the first projection is a homeomorphism). It will be useful, however, to think of it as being obtained by endowing two copies of $\mathbb{C} \smallsetminus 0$, which we call X^+ and

X^-, with a nonstandard topology, as illustrated in Figure 1.6. We note before-hand that intuitively this topology amounts to having slit X^- and X^+ along the positive real half-lines and reglued in the opposite way the resulting "loose ends".

Denote $B_r^+(z)$ (respectively $B_r^-(z)$) a Euclidean open ball in X^+ (respectively X^-) centered at z with radius r. Endow $X^+ \cup X^-$ with topology generated by the following basis:

- Euclidean open balls in X^+ or X^- which do not intersect the half-line of positive real numbers.
- For $0 < \epsilon < x \in \mathbb{R}$,

$$\left(B_\epsilon^+(x) \cap \{Im(z) \geq 0\}\right) \cup \left(B_\epsilon^-(x) \cap \{Im(z) < 0\}\right)$$

and

$$\left(B_\epsilon^-(x) \cap \{Im(z) \geq 0\}\right) \cup \left(B_\epsilon^+(x) \cap \{Im(z) < 0\}\right).$$

- For $0 < r_1, r_2 \in \mathbb{R}$,

$$B_{r_1}^-(0) \cup B_{r_2}^+(0).$$

Exercise 1.4.3. Using the above construction of $\Gamma_2 \setminus (0,0) = X^+ \cup X^-$, understand how this space acts as a domain for the inverse map $z = w^{1/2}$. For example, follow the images $w^{1/2}$ as you loop around the origin again and again in $X^+ \cup X^-$.

One should ignore the fact that the end result of this construction is yet again just a copy of $\mathbb{C} \setminus 0$, and rather think of the fact that by "slitting and regluing" two copies of $\mathbb{C} \setminus 0$ you are constructing a new space which around every point is indistinguishable from the complex numbers. Such a space is some kind of "escalator" that allows you to connect the various branches of local inverses of f around the critical point.

The reason for this alternative, and more complicated, point of view is that the ideas of the construction can be applied in general to give a local model for the inverse of any holomorphic function f in a neighborhood of a critical point, i.e. a point where the derivative of f vanishes (this statement is made precise in Lemma 1.4.4).

Exercise 1.4.4. Let $g(w)$ be a holomorphic function from $\mathbb{C}$ to $\mathbb{C}$; show that, for any w_0 such that $g(w_0) \neq 0$, there exists a neighborhood of w_0, and k distinct choices (called branches) for a holomorphic map $\tilde{g}$ such that $\tilde{g}^k(w) =$

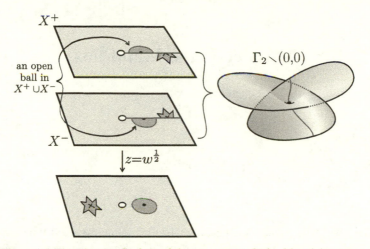

Figure 1.6 The Riemann Surface of the square root obtained by giving a non-standard topology to two copies of $\mathbb{C} \smallsetminus (0, 0)$. The shaded areas represent open sets in $X^{-} \cup X^{+}$.

$g(w)$. (When we don't care particularly about which branch one may choose, this function is (with abuse of notation) denoted by $g(w)^{1/k}$ or $\sqrt[k]{g(w)}$.)

Lemma 1.4.4. *Let $w = f(z)$ be a holomorphic function and $z_0 \in \mathbb{C}$ such that the first $k-1$ derivatives of $f(z)$ vanish at z_0 but $f^{(k)}(z_0) \neq 0$. Then there exist holomorphic changes of variable $\tilde{z}(z)$, $\tilde{w}(w)$ such that in the new variables f becomes $\tilde{w} = \tilde{z}^k$.*

Proof Consider the Taylor expansion of f around z_0. By the hypothesis that all derivatives of order less than k vanish, it must be that:

$$f(z) - f(z_0) = \sum_{n=k}^{\infty} a_n(z - z_0)^n, \qquad (1.10)$$

and since $f^{(k)}(z_0) \neq 0$, it must be that $a_k \neq 0$. The function $g(z) = \sum_{n=k}^{\infty} a_n(z - z_0)^{n-k}$ is holomorphic and such that $g(z_0) \neq 0$; therefore it admits a branch of the k-th root around z_0.

We now define $\tilde{z} = (z - z_0)\sqrt[k]{g(z)}$. This is naturally a holomorphic function, and one can show that it is invertible at z_0 (and hence a legitimate change of variable) by showing that $\frac{d}{dz}\tilde{z}_{|z_0} \neq 0$. We leave this check as an exercise. Finally, defining $\tilde{w} = w - f(z_0)$, we note that (1.10) is expressed in the new variables as $\tilde{w} = \tilde{z}^k$. $\square$

2

Introduction to Manifolds

In Chapter 1 we created an honest domain for the multi-valued function $z^{1/2}$ by "gluing together" open subsets of $\mathbb{C}$. This construction generalizes to the notion of a *manifold*: a space whose global geometry may be complicated, but such that the local geometry around any point is familiar. An illustration to have in mind is how the Earth can locally be represented on your flat computer screen (i.e. an open set in $\mathbb{R}^2$) in Google Maps, but globally the Earth is – *spoiler alert!* – roughly spherical.

Suppose you are Google-mapping your neighborhood with your house at the center of your screen. We think of the map you are looking at as a function

$$\varphi : \text{(a subset of the Earth)} \rightarrow \text{(your flat computer screen)}.$$

Such a function, which we will call a *chart*, gives coordinates to a little piece of Earth simply by adopting, for that piece of Earth, the coordinates of your computer screen.

Now imagine that your friend who lives down the street is also viewing a map of your neighborhood, centered on his house. Your friend's screen doesn't look exactly the same as yours, but there is a portion of your neighborhood which is on both your screen and your friend's screen. There is a natural function

$$T : \left(\begin{array}{c} \text{the common portion of your} \\ \text{neighborhood on your screen} \end{array} \right) \rightarrow \left(\begin{array}{c} \text{the common portion on your} \\ \text{friend's screen} \end{array} \right)$$

which identifies the points on the two screens that correspond to the same physical location. By using this function, you and your friend can piece together your neighborhood and, for example, figure out how to get from a location that appears on your screen but not on his to a location that appears on his screen but not on yours.

The function T will be called a *transition function*. The role of transition functions is to compare geometric information coming from different charts.

The key philosophical idea is, then, that we understand completely the geometry of the Earth through the following information:

- a set of charts, the union of whose domains covers the Earth;
- transition functions that allow us to compare information among charts whose domains overlap.

In the olden days before the internet, such information was collected in big books, often called *World Atlases*, in which each page corresponds to a chart and the transition functions are provided by the little strips on the boundary of each page that are repeated in some other page.

2.1 General Definition of a Manifold

Now for a formal definition of a manifold. We invite the reader to refer to Figure 2.1.

Definition 2.1.1. A topological space X is called a **[smooth] manifold** if and only if the following conditions are satisfied.

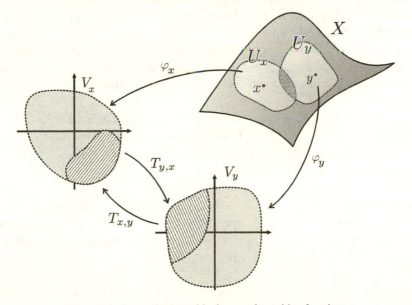

Figure 2.1 A manifold X with charts and transition functions

1. X is a Hausdorff topological space;
2. For all $x \in X$ there exists a neighborhood $U_x \subset X$ of x and a homeomorphism $\varphi_x : U_x \to V_x$ where V_x is an open set in $[\mathbb{R}^n]$;
3. For any U_x, U_y such that $U_x \cap U_y \neq \varnothing$ the **transition function**

$$T_{y,x} : \varphi_y \circ \varphi_x^{-1} : \varphi_x(U_x \cap U_y) \to \varphi_y(U_x \cap U_y)$$

is [smooth].

Recall that a function $\vec{f} = (f_1, f_2, \ldots, f_n) : \mathbb{R}^m \to \mathbb{R}^n$ is **smooth**, i.e. $f \in C^\infty$, if each f_i has continuous partial derivatives of all orders; in other words, if all partial derivatives $\partial^k f_i / \partial x_{i_1} \partial x_{i_2} \cdots \partial x_{i_k}$ exist and are continuous.

Remark 2.1.2. We have placed (purely notational) brackets throughout Definition 2.1.1 to indicate that there is a class of manifold for many different "categories" relating to the real and complex numbers. For instance, one can consider differentiable C^k-manifolds, or complex analytic manifolds. In the latter case, we have $V_x \subset \mathbb{C}^n$ and we ask transition functions to be holomorphic. This will be our focus from Chapter 3 onward; however, smooth manifolds lend themselves well to visualization and so are well-suited to an introduction.

Definition 2.1.3. The n in $\mathbb{R}^n$ (or $\mathbb{C}^n$ if considering complex analytic manifolds) in part 2 of Definition 2.1.1 is called the **dimension** of X. The pair (U_x, φ_x) is called a **local chart** for X, and the function φ_x is called a **local coordinate function**. A transition function compares different local coordinates for the same points of X and is therefore also called a **change of coordinates**. A collection $\mathcal{A} = \{(U_\alpha, \varphi_\alpha)\}_\alpha$ of local charts that covers X, such that all transition functions are [smooth], is called an **atlas**.

The same topological space can be given the structure of a manifold in different ways, i.e. can be given different atlases. However, some atlases determine the same manifold structure on the topological space; when this happens we say the two atlases are compatible.

Definition 2.1.4. Two atlases $\mathcal{A} = \{(U_\alpha, \varphi_\alpha)\}_\alpha$ and $\mathcal{B} = \{(U_\beta, \varphi_\beta)\}_\beta$ for a topological space X are called **compatible** if their union $\mathcal{A} \cup \mathcal{B}$ is an atlas for X; in other words, if for all α, β such that $U_\alpha \cap U_\beta \neq \varnothing$ the transition functions $\varphi_\beta \circ \varphi_\alpha^{-1}$ and $\varphi_\alpha \circ \varphi_\beta^{-1}$ are [smooth].

Exercise 2.1.1. Show that compatibility is an equivalence relation on the collection of atlases for a topological space X.

An equivalence class of compatible atlases for X is called a [smooth differentiable] **structure** on X.

2.2 Basic Examples

The most trivial example of a smooth manifold is Euclidean space itself. In this case $\mathbb{R}^n$ can be covered by the one chart atlas $(\mathbb{R}^n, Id)$, and there are no transition functions to worry about. Similarly, any open subset of $\mathbb{R}^n$ is a smooth manifold.

Example 2.2.1. We give the unit circle $S^1 = \{(x, y) \in \mathbb{R}^2 | x^2 + y^2 = 1\}$ with topology induced by $\mathbb{R}^2$ the structure of a smooth manifold. Note that S^1 is Hausdorff since $\mathbb{R}^2$ is. We define an atlas (illustrated in Figure 2.2) consisting of four charts with the following domains:

- $U_x^+ = \{(x, y) \in S^1 | x > 0\}$
- $U_x^- = \{(x, y) \in S^1 | x < 0\}$
- $U_y^+ = \{(x, y) \in S^1 | y > 0\}$
- $U_y^- = \{(x, y) \in S^1 | y < 0\}$.

We use projection to define our coordinate functions:

- $\varphi_x^\pm = \pi_y|_{U_x^\pm} : U_x^\pm \to (-1, 1) \subset \mathbb{R}$ sends $(x, y) \mapsto y$
- $\varphi_y^\pm = \pi_x|_{U_y^\pm} : U_y^\pm \to (-1, 1) \subset \mathbb{R}$ sends $(x, y) \mapsto x$.

Note that $(\varphi_y^+)^{-1}(x) = (x, \sqrt{1 - x^2})$ is a continuous function on the interval $(-1, 1)$ and hence φ_y^+ is a homeomorphism. Similarly, one can check that all other local coordinate functions are homeomorphisms.

Consider the intersection $U := U_x^+ \cap U_y^+ = \{(x, y) \in S^1 | x, y > 0\}$; we have $\varphi_x^+(U) = \varphi_y^+(U) = (0, 1) \subset \mathbb{R}$. The transition function $T_{x+, y+} = \varphi_x^+ \circ (\varphi_y^+)^{-1}$ sends $x \in (0, 1)$ to

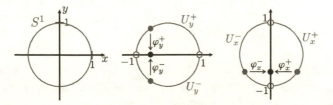

Figure 2.2 A circle with four charts

$$T_{x^+,y^+} : x \xrightarrow{(\varphi_y^+)^{-1}} (x, \sqrt{1-x^2}) \xrightarrow{\varphi_x^+} \sqrt{1-x^2},$$

which is smooth on the domain $(0, 1)$. Similar checks show that all other transition functions are smooth. Thus our atlas gives S^1 the structure of a smooth manifold.

Exercise 2.2.1. Another method of giving S^1 a manifold structure is by **stereographic projection**. Define the points $N = (0, 1)$, $S = (0, -1)$ and corresponding subsets $U_N = S^1 - N$, $U_S = S^1 - S$. Define the chart $\varphi_N : U_N \to \mathbb{R}$ by $\varphi_N(x_0, y_0) = x_N$, where x_N is the x-intercept of the unique line through N and (x_0, y_0) in $\mathbb{R}^2$. Define φ_S by replacing N with S.

1. Find explicit formulas for $\varphi_N(x_0, y_0)$, $\varphi_S(x_0, y_0)$ in terms of x_0 and y_0.
2. Let $x_N \in \mathbb{R}$. Show that the x-coordinate of $(\varphi_N)^{-1}(x_N)$ is $\frac{2x_N}{x_N^2+1}$ and that the y-coordinate is the following:

$$\begin{cases} \sqrt{1 - (\frac{2x_N}{x_N^2+1})^2} & \text{if } |x_N| \geq 1 \\ -\sqrt{1 - (\frac{2x_N}{x_N^2+1})^2} & \text{if } |x_N| < 1. \end{cases}$$

3. Consider the transition function $T_{S,N} = \varphi_S \circ (\varphi_N)^{-1}$.
   (a) What are the domain and range of $T_{S,N}$?
   (b) Compute $T_{S,N}$ in terms of x_N. *Hint: there will be two cases.*
   (c) Show that $T_{S,N}$ is a smooth function on its domain.
   Note that the check for the smoothness of $T_{N,S}$ is completely analogous.
4. Show that the atlas $\mathcal{B}$ given here using stereographic projection and the atlas $\mathcal{A}$ defined in Example 2.2.1 are compatible. For any local chart φ_α from $\mathcal{A}$ and any local chart φ_β from $\mathcal{B}$ whose domains intersect, you should compute $\varphi_\beta \circ \varphi_\alpha^{-1}$ and $\varphi_\alpha \circ \varphi_\beta^{-1}$ and show that they are smooth functions on their domains.

Exercise 2.2.2. The two atlases considered for S^1 in Example 2.2.1 and Exercise 2.2.1 generalize to give atlases for each S^n, $n \geq 1$. In dimension n, the atlas analogous to $\mathcal{A}$ has $2n + 2$ charts and the one analogous to $\mathcal{B}$ still has two charts. Understand these generalizations.

Example 2.2.2. Let the function f map $\mathcal{M}(m, n, \mathbb{R}) = \{m \times n$ matrices with real entries$\}$ to $\mathbb{R}^{mn}$ by fixing once and for all a bijection between the mn positions in a matrix and the coordinates of $\mathbb{R}^{mn}$ and sending each entry to the corresponding coordinate. For example, we have

$$f : \begin{pmatrix} 1 & 2 & 3 \\ 4 & 5 & 6 \end{pmatrix} \in \mathcal{M}(2, 3, \mathbb{R}) \longmapsto (1, 2, 3, 4, 5, 6) \in \mathbb{R}^6.$$

We induce a topology on $\mathcal{M}(m, n, \mathbb{R})$ by *defining* f to be a homeomorphism, i.e. $U \subset \mathcal{M}(m, n, \mathbb{R})$ is open if and only if $f(U)$ is open in $\mathbb{R}^{mn}$. Then the one-chart atlas $\{\mathcal{M}(m, n, \mathbb{R}), f\}$ gives $\mathcal{M}(m, n, \mathbb{R})$ the structure of a smooth manifold.

Exercise 2.2.3. Show that any open set of a smooth manifold X is itself a smooth manifold.

Example 2.2.3. The set $GL(n, \mathbb{R}) = \{M \in \mathcal{M}(n, n, \mathbb{R}) | \det(M) \neq 0\}$ is a smooth manifold since it is an open set of a smooth manifold.

2.3 Projective Spaces

Projective spaces are important and "historical" examples of manifolds; even though the current formalization of the theory happened much later, ideas in projective geometry date back to Pappus (290–350 AD). Renaissance painters Leon Battista Alberti (1404–72) and Piero della Francesca (1410–92) wrote mathematical treatises on planar projective geometry arising from their studies of perspective drawing. The basic idea is the following: a painter represents the three-dimensional world by projecting it onto a two-dimensional canvas. All points that lie on the same line through the eye of the painter end up at the same point on the canvas (see Figure 2.3).

Projective spaces capture and formalize this idea mathematically: they are geometric objects whose points are in bijection with lines through the origin

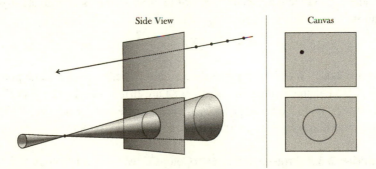

Figure 2.3 Points in space lying on the same "line of sight" get drawn as one point on the artist's canvas

(the painter's eye) in an ambient Euclidean space. Natural charts that define a manifold structure are given by all possible canvases that the painter may place in space (away from his/her eye) and project onto. We now define projective space in three stages: first we describe the set of points, then the topology and finally the manifold structure.

Definition 2.3.1 (Projective Space: Points). The **set of points** of $\mathbb{P}^n(\mathbb{R})$ is defined to be naturally in bijection with either of the following sets:

1. lines ℓ through the origin in $\mathbb{R}^{n+1}$;
2. equivalence classes of $n + 1$-tuples of real numbers $(X_0, \ldots, X_n) \neq (0, \ldots, 0)$ such that, for any $\lambda \in \mathbb{R} \smallsetminus \{0\}$,

$$(X_0, \ldots, X_n) \sim (\lambda X_0, \ldots, \lambda X_n).$$

Remark 2.3.2. For the reader familiar with coordinate-free linear algebra, it is worth pointing out that, given an $n + 1$ dimensional $\mathbb{R}$-vector space V, one can define the **projectivization of** V (denoted $\mathbb{P}(V)$), whose points correspond to one-dimensional linear subspaces of V.

Exercise 2.3.1. Show that the two sets in Definition 2.3.1 are canonically in bijection with each other, and either is put in bijection with the set in Remark 2.3.2 by choosing a basis for V. Therefore, any one of them can be taken as a model for the points of $\mathbb{P}^{n-1}(\mathbb{R})$.

We can use $n + 1$-tuples of numbers to identify points of $\mathbb{P}^n(\mathbb{R})$, much like coordinates in a vector space after choosing a basis. However, we require coordinates for a point to be unique, and this is not the case here. The $n + 1$-tuples $(X_0, \ldots, X_n)$ are called **homogeneous coordinates**; given a point $\ell \in \mathbb{P}^n(\mathbb{R})$, we denote it via the equivalence class of its homogeneous coordinates by the notation

$$\ell = [X_0 : X_1 : \ldots : X_n].$$

Exercise 2.3.2. Which of the points listed below represent the same point in $\mathbb{P}^1(\mathbb{R})$? How many distinct points of $\mathbb{P}^1(\mathbb{R})$ are listed? $[1 : 1]$, $[2 : -1/2]$, $[0 : 1]$, $[-1/4 : -1/4]$, $[6 : -3/2]$, $[-2 : -2]$. As a side puzzle: why is $[0 : 0]$ not a legitimate homogeneous coordinate for any point in $\mathbb{P}^1(\mathbb{R})$?

Definition 2.3.3 (Projective Space: Topology). We give a topology to $\mathbb{P}^n(\mathbb{R})$ by inducing it as the quotient topology via a surjective function. Consider the natural projection function:

$$\pi : \quad \mathbb{R}^{n+1} \smallsetminus \{\vec{0}\} \quad \rightarrow \quad \mathbb{P}^n(\mathbb{R})$$
$$(X_0, X_1, \ldots, X_n) \quad \mapsto \quad [X_0 : X_1 : \ldots : X_n].$$

A set $U \subseteq \mathbb{P}^n(\mathbb{R})$ is defined to be open if and only if $\pi^{-1}(U)$ is open in $\mathbb{R}^{n+1} \smallsetminus \{0\}$. In other words, we give $\mathbb{P}^n(\mathbb{R})$ the finest topology that makes π continuous.

Exercise 2.3.3. In Definition 2.3.3 we realized $\mathbb{P}^n(\mathbb{R})$ as an identification/orbit space: let $\mathbb{R}^* = \mathbb{R} \smallsetminus \{0\}$ act on $\mathbb{R}^{n+1}$ by component-wise multiplication: $\lambda \cdot (X_0, X_1, \ldots, X_n) = (\lambda X_0, \lambda X_1, \ldots, \lambda X_N)$. Then

$$\mathbb{P}^n(\mathbb{R}) = \left(\mathbb{R}^{n+1} \smallsetminus \{\vec{0}\}\right) / \mathbb{R}^*.$$

We now present two alternative models for $\mathbb{P}^n(\mathbb{R})$ as an identification space, and leave it as an exercise that they yield homeomorphic results.

Sphere quotient. Consider the n-dimensional unit sphere $S^n \subset \mathbb{R}^{n+1}$. The multiplicative cyclic group $\mu_2 = \{1, -1\}$ acts on the sphere by

$$\pm 1 \cdot (X_0, X_1, \ldots, X_n) = (\pm X_0, \pm X_1, \ldots, \pm X_n).$$

Then $\mathbb{P}^n(\mathbb{R})$ is the quotient space S^n/μ_2.

Disk model. Consider the n-dimensional closed unit disk $\overline{D}^n \subset \mathbb{R}^n$, and consider the antipodal equivalence relation on the points of its boundary: $\mathbf{x} \sim -\mathbf{x}$ if and only if $||\mathbf{x}|| = 1$. Then $\mathbb{P}^n(\mathbb{R})$ is the identification space $\overline{D}^n/\sim$.

We point out that the sphere quotient and disk model for $\mathbb{P}^n(\mathbb{R})$ immediately show that projective space is compact, and we make it an exercise to show it is Haussdorf.

We now give real projective space the structure of a smooth manifold. In order to avoid clouding the ideas with cumbersome notation, we treat explicitly the case $n = 1$ and leave it to the reader to draw the natural generalizations.

We describe explicitly a two-chart atlas defining a smooth manifold structure on $\mathbb{P}^1(\mathbb{R})$.

Inside $\mathbb{R}^2$ with coordinates (X, Y), identify the line $\{X = 1\}$ with $\mathbb{R}$ by using $y = Y$ as a coordinate. Each non-vertical line intersects the line $\{X = 1\}$ at a unique point, and this association determines our coordinate function. Formally, we define

$$U_X = \mathbb{P}^1(\mathbb{R}) \smallsetminus \{[0 : 1]\} = \{[X : Y] \in \mathbb{P}^1(\mathbb{R}) | X \neq 0\}$$

and $\varphi_X : U_X \to \mathbb{C}$ by

$$\varphi_X([X : Y]) = Y/X = y$$

(note that $X \neq 0$ implies $[X : Y] = [1 : Y/X]$).

Similarly, we define a second chart using the line $\{Y = 1\}$, i.e.

$$U_Y = \mathbb{P}^1(\mathbb{R}) \setminus \{[1 : 0]\} = \{[X : Y] \in \mathbb{P}^1(\mathbb{R}) | Y \neq 0\}$$

and $\varphi_Y : U_Y \to \mathbb{C}$

$$\varphi_Y([X : Y]) = X/Y = x.$$

Exercise 2.3.4. Draw a picture illustrating the coordinate functions φ_X and φ_Y. Show that φ_X and φ_Y are homeomorphisms.

We now consider transition functions. We have $U := U_X \cap U_Y = \{[X : Y] \in \mathbb{P}^1(\mathbb{R}) | X, Y \neq 0\}$ and $\varphi_X(U) = \varphi_Y(U) = \mathbb{R} \setminus \{0\}$. The transition function $T_{x,y} = \varphi_Y \circ (\varphi_X)^{-1}$ sends $y \neq 0$ to

$$T_{x,y} : y \xrightarrow{(\varphi_x)^{-1}} [1 : y] = [1/y : 1] \xrightarrow{\varphi_y} 1/y$$

which is smooth on the domain $\mathbb{R} \setminus \{0\}$. Similarly $T_{x,y} : x \mapsto 1/x$ is smooth, and thus $\mathbb{P}^1(\mathbb{R})$ is a smooth manifold.

Exercise 2.3.5. Convince yourself that $\mathbb{P}^1(\mathbb{R})$ is homeomorphic to a circle. Since $\mathbb{P}^1(\mathbb{R})$ has dimension 1, it is called the **projective line**.

Remark 2.3.4. The charts U_X and U_Y are often called **coordinate affine charts** and the functions x and y **affine coordinates**. Since, in the case of the projective line, any one affine chart captures all of the space except one point, it is common to describe the space with just one affine coordinate, and allow it to take the value ∞ to represent the missing point. Note, however, that the notion of a point being at infinity depends on the choice of the affine chart that is being used. When using U_X with affine coordinate y, then $y = \infty$ corresponds to the point $[0 : 1]$, whereas for U_Y the point $x = \infty$ is $[1 : 0]$.

Exercise 2.3.6. Our choice of atlas for $\mathbb{P}^1(\mathbb{R})$ is somewhat arbitrary (at least mathematically). Show that any choice of two non-parallel lines not through the origin gives an atlas compatible with our choice.

Exercise 2.3.7. Show that the following three charts form an atlas for the **real projective plane**, $\mathbb{P}^2(\mathbb{R})$:

$$U_X = \{X \neq 0\}, \quad \varphi_X([X : Y : Z]) = (Y/X, Z/X) \tag{2.1}$$

$$U_Y = \{Y \neq 0\}, \quad \varphi_Y([X : Y : Z]) = (X/Y, Z/Y) \tag{2.2}$$

$$U_Z = \{Z \neq 0\}, \quad \varphi_Z([X : Y : Z]) = (X/Z, Y/Z). \tag{2.3}$$

Generalize this construction to show that $\mathbb{P}^n(\mathbb{R})$ is a smooth manifold for any positive integer n.

Giving the structure of a complex analytic manifold to the complex projective spaces $\mathbb{P}^n(\mathbb{C})$ might seem trickier, as one has to consider complex lines through the origin in $\mathbb{C}^{n+1}$, but the construction of the local charts and checks of transition functions follow completely in analogy to the real case.

2.4 Compact Surfaces

This paragraph is an essential introduction to surfaces, the manifolds we really care about in our story. We recall the theorem of classification of compact surfaces, present the notion of an identification polygon representing a surface and introduce Euler characteristic and orientability, the two fundamental topological invariants for surfaces. The reader interested in a complete account and rigorous proofs may consult any basic topology textbook; for example, Armstrong (1983) and Munkres (1975).

Definition 2.4.1. A **surface** is a manifold of real dimension 2.

Trivial examples of surfaces are given by $\mathbb{R}^2, \mathbb{C}$ and any of their open subsets. We are especially interested in connected, compact surfaces, i.e. surfaces that are connected and compact as topological spaces. In Section 2.2 we showed that the two-dimensional sphere S^2 is an example of a compact surface, and in Section 2.3 we introduced the more exotic example $\mathbb{P}^2(\mathbb{R})$. A third example that may be familiar is the **torus** T, corresponding to the "glaze" of a doughnut. A remarkable theorem in topology tells us that these three surfaces may be used as building blocks to construct every other connected, compact surface.

Definition 2.4.2. Given two connected surfaces S_1 and S_2, the **connected sum** $S_1 \# S_2$ is the surface obtained by removing an open disk from each of the surfaces and identifying the resulting boundaries via a homeomorphism.

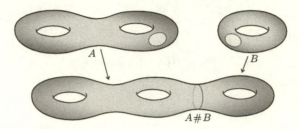

Figure 2.4 The connected sum of two surfaces

The operation of connected sum is illustrated in Figure 2.4. Of course, one must prove that this operation is well defined up to homeomorphism, i.e. the result is independent of the choice of disks to remove and of the homeomorphism used to identify the boundaries.

Exercise 2.4.1. Assuming that the operation of connected sum is well defined, show that it gives the structure of an associative monoid with identity to the set of homeomorphism classes of connected compact surfaces. Which surface is the identity element?

The theorem of classification of compact surfaces states that, up to homeomorphism, there is the sphere plus two countable classes of connected, compact surfaces.

Theorem 2.4.3 (Classification of Compact Surfaces). *Any connected, compact surface is homeomorphic to exactly one surface in the following list. The indices g, m take value among all positive integers:*

- S^2, *the two-dimensional sphere;*
- $T^{\#g} = T\#\ldots\#T$, *the connected sum of g tori;*
- $\mathbb{P}^2(\mathbb{R})^{\#m} = \mathbb{P}^2(\mathbb{R})\#\ldots\#\mathbb{P}^2(\mathbb{R})$, *the connected sum of m projective planes.*

A complete proof of this theorem would be too large a detour from our path, but the one-paragraph sketch goes as follows: to show that the above list is exhaustive, one shows that any surface may be represented by an *identification polygon*, and then show via an inductive procedure that all identification polygons give surfaces in the above list. To show that no two distinct surfaces in the above list are homeomorphic, one constructs two topological invariants, *orientability* and *Euler characteristic*, and shows that they are a complete set of invariants for the above list of surfaces, meaning that no two distinct elements take the same value for both invariants. In the remaining part of this paragraph we introduce these notions.

2.4.1 Identification Polygons

First for some abstract nonsense: we call a set A of n letters an alphabet, and we call the set $A \cup \overline{A}$, consisting of repeating each letter a second time with a bar above it, a doubled alphabet. Each pair $a, \bar{a}$, for $a \in A$, is called a pair of twin letters.

Definition 2.4.4. An **identification polygon with** $2n$ **sides** is a word w constructed from a doubled n-letter alphabet such that, for each pair of twin letters, w contains exactly two letters from that pair (repetitions allowed). In particular, the word w must have exactly $2n$ letters.

Example 2.4.5. An alphabet of two letters consists of $A = \{a, b\}$. Then the doubled alphabet is $A \cup \overline{A} = \{a, b, \bar{a}, \bar{b}\}$ and the twin pairs are $\{a, \bar{a}\}$ and $\{b, \bar{b}\}$. Then the following are examples of identification polygons:

$$w_1 = a\bar{a}b\bar{b} \quad w_2 = aab\bar{b} \quad w_3 = ab\bar{a}\bar{b} \quad w_4 = abab.$$

An identification polygon w gives rise to a compact surface as follows. Consider a regular $2n$-gon and label its sides, counterclockwise, so as to spell the word w. Then, for every twin pair in the alphabet, identify the two sides of the polygon labeled by that twin pair via a homeomorphism which preserves the orientation of the boundary if the two letters are the same, and via a homeomorphism which reverses the orientation of the boundary if both elements in the twin pair appear in the labeling. Any surface S homeomorphic to the surface thus obtained is said to be **represented** by the identification polygon w.

Example 2.4.6. The identification polygon $w = a\bar{a}$ gives a surface homeomorphic to the sphere S^2, as shown in Figure 2.5. The polygon aa is precisely the disk model for the real projective plane $\mathbb{P}^2(\mathbb{R})$ (as in Section 2.3).

Exercise 2.4.2. Recognize the surfaces of the four identification polygons in Example 2.4.5 as the sphere, the real projective plane, the torus and the Klein bottle.

Identification polygons play very well with connected sums, as we explore in the next exercise.

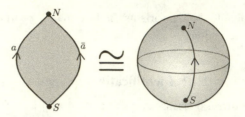

Figure 2.5 The identification polygon $w = a\bar{a}$ gives a sphere S^2: the boundary of w is like a "zipper", and the identification process consists in "zipping it up"

Exercise 2.4.3. If S_1, S_2 are connected compact surfaces represented by the polygons w_1 and w_2, then show that the surface $S_1\#S_2$ is represented by the polygon w_1w_2 (i.e. the concatenation of the two words). Note that we are assuming that the two words use letters from different alphabets, A_1 and A_2.

If you are having fun with this whirlwind incursion into identification polygons, you may want to challenge yourself with the following exercises.

Exercise 2.4.4. By appropriately manipulating identification polygons, show the following homeomorphisms:

- Klein bottle $\cong \mathbb{P}^2(\mathbb{R})\#\mathbb{P}^2(\mathbb{R})$
- $T\#\mathbb{P}^2(\mathbb{R}) \cong \mathbb{P}^2(\mathbb{R})\#\mathbb{P}^2(\mathbb{R})\#\mathbb{P}^2(\mathbb{R})$.

Exercise 2.4.5. Show that any identification polygon gives a surface homeomorphic to one in the list appearing in Theorem 2.4.3.

2.4.2 Euler Characteristic and Orientability

The Euler characteristic of a surface is a number computed through an auxiliary "good" graph on the surface.

Definition 2.4.7. A **good graph** on a surface S is a graph Γ on S such that:

1. $S \smallsetminus \Gamma$ is homeomorphic to a disjoint union of open disks;
2. wherever two edges cross there is a vertex;
3. no edge ends without a vertex.

The **Euler Characteristic** of a surface S is a topological invariant which can be defined/computed as:

$$\chi(S) = |V_\Gamma| - |E_\Gamma| + |F_\Gamma|,$$

where V_Γ, E_Γ, F_Γ are the sets of vertices, edges and faces of a "good graph".

The Euler characteristic is in fact independent of the choice of a good graph on S, and is therefore a topological invariant.

Example 2.4.8. Any platonic solid can be thought of as a topological sphere together with a good graph on it: vertices, edges and faces are simply those of the platonic solid. For example, a tetrahedron has 4 vertices, 6 edges and 4 faces; a cube gives 8 vertices, 12 edges and 6 faces. The Euler characteristic of the sphere is thus computed to be $\chi(S^2) = 2$.

Exercise 2.4.6. Show that if S is represented by an identification polygon w, then the boundary of the polygon w is a good graph on S.

Exercise 2.4.7. If S_1 and S_2 are compact, connected surfaces:

$$\chi(S_1 \# S_2) = \chi(S_1) + \chi(S_2) - 2.$$

Exercise 2.4.8. Compute:

- $\chi(T^{\#g}) = 2 - 2g$;
- $\chi(\mathbb{P}^2(\mathbb{R})^{\#m}) = 2 - m$.

By using the standard representation of a compact surface as an identification polygon, one can deduce that the Euler characteristic of a compact orientable surface of genus g is $2 - 2g$.

Orientability is a somewhat more sophisticated invariant.

Definition 2.4.9. A surface S is **orientable** if it admits an atlas such that all transition functions are orientation-preserving (in the sense of Corollary 1.1.3). Such an atlas is called a **positive atlas** for S.

The prototypical example of a non-orientable surface is the Möbius strip, depicted in Figure 2.6. In fact, one can show that a surface is non-orientable if and only if it contains an open subset homeomorphic to a Möbius strip.

Exercise 2.4.9. Show that for all $m \geq 1$, $\mathbb{P}^2(\mathbb{R})^{\#m}$ are non-orientable.

Fact: The sphere, the torus and all connected sums of tori are orientable surfaces.

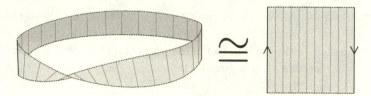

Figure 2.6 A Möbius strip

Exercise 2.4.10. Show that Euler characteristic and orientability are a complete set of invariants for compact, connected surfaces.

2.5 Manifolds as Level Sets

A natural way to construct interesting manifolds as subsets of Euclidean space is by considering level sets of functions. Given the function $f(x, y) = x^2 + y^2$, the unit circle can be thought of as the level set $f^{-1}(1)$; in Example 2.2.1 we showed that the circle is a manifold.

Let us now consider instead the function $g(x, y) = xy$. The level set $g^{-1}(1)$ is a smooth hyperbola, which you may check is a one-dimensional smooth manifold. But the level set $g^{-1}(0)$ consists of the union of the x and y axes, which cannot be given the structure of a smooth manifold: no neighborhood of the origin can be homeomorphic to an open set in $\mathbb{R}$.

The answer to which level sets are well behaved lies in this classical theorem from analysis.

Theorem 2.5.1 (The Implicit Function Theorem). *Let* $F : \mathbb{R}^n \to \mathbb{R}^m$ *be a smooth function, and* $\mathbf{x} \in \mathbb{R}^n$ *such that the differential* $dF(\mathbf{x})$ *is a surjective linear function. Say* $F(\mathbf{x}) = \mathbf{a}$. *Then there exist:*

- $V_{\mathbf{x}} \subseteq \mathbb{R}^n$, *an open neighborhood of* $\mathbf{x}$;
- $U_{\mathbf{x}} \subseteq \mathbb{R}^{n-m}$ *an open set;*
- $f_{\mathbf{x}} : U_{\mathbf{x}} \to \mathbb{R}^m$ *a smooth function*

such that

$$F^{-1}(\mathbf{a}) \cap V_{\mathbf{x}} = \Gamma_f,$$

where Γ_f *denotes the graph of* f.

Proof The geometric idea for the proof of this theorem is illustrated in Figure 2.7, which we invite the reader to refer to throughout the proof; to

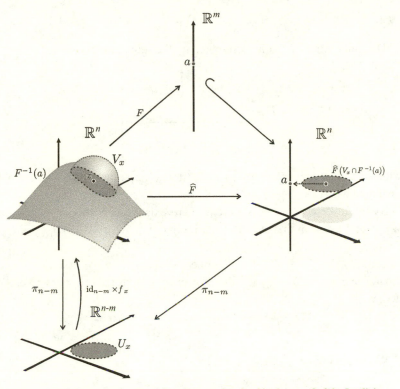

Figure 2.7 Schematic picture of the geometry involved in the proof of the Implicit Function Theorem

do things honestly we must alas use coordinates – let us bite the bullet and do it.

Let $F = (F_1(x_1, \ldots, x_n), \ldots, F_m(x_1, \ldots, x_n))$, and consider the $m \times n$ matrix of partial derivatives:

$$J = \begin{bmatrix} \frac{\partial F_1}{\partial x_1}(\mathbf{x}) & \cdots & \frac{\partial F_1}{\partial x_n}(\mathbf{x}) \\ \vdots & \ddots & \vdots \\ \frac{\partial F_m}{\partial x_1}(\mathbf{x}) & \cdots & \frac{\partial F_m}{\partial x_n}(\mathbf{x}) \end{bmatrix}$$

The differential $dF(\mathbf{x})$ is the linear function represented by the matrix J, and therefore it is surjective if and only if $n \geq m$ and J has maximal rank m. We assume without loss of generality that the rightmost $m \times m$ minor of J has nonzero determinant.

We now consider an auxiliary function $\hat{F} : \mathbb{R}^n \to \mathbb{R}^n$, specifically constructed so that it extends F and satisfies the hypotheses of the Inverse Function Theorem (1.4.1):

$$\hat{F}(x_1, \ldots, x_n) = (x_1, x_2, \ldots, x_{n-m}, F_1(x_1, \ldots, x_n), \ldots, F_m(x_1, \ldots, x_n)).$$

The differential $d\hat{F}(\mathbf{x})$ is represented by the square matrix

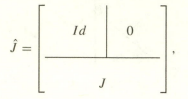

which has determinant equal to the last $m \times m$ minor of J, nonzero by assumption.

By the Inverse Function Theorem, there exist neighborhoods $V_{\mathbf{x}}$ of $\mathbf{x}$ and $W_{\hat{F}(\mathbf{x})}$ of $\hat{F}(\mathbf{x})$ such that smooth local inverse function $\hat{F}^{-1} : W_{\hat{F}(\mathbf{x})} \to U_{\mathbf{x}}$ is defined.

Let $\pi_{n-m} : \mathbb{R}^n \to \mathbb{R}^{n-m}$ be the linear projection on the first $n - m$ coordinates and $\pi_m : \mathbb{R}^n \to \mathbb{R}^m$ the projection on the last m coordinates. We define

$$U_{\mathbf{x}} = \pi_{n-m}\left(W_{\hat{F}(\mathbf{x})} \cap \{x_{n-m+1} = a_1\} \ldots \cap \{x_n = a_m\}\right).$$

We finally define: $f_{\mathbf{x}} : U_{\mathbf{x}} \to \mathbb{R}^m$ by

$$f_{\mathbf{x}}(x_1, \ldots, x_{n-m}) = \pi_m(\hat{F}^{-1}(x_1, \ldots, x_{n-m}, a_1, \ldots, a_m)).$$

Checking that $V_{\mathbf{x}}, U_{\mathbf{x}}$ and $f_{\mathbf{x}}$ verify the statement of the theorem is now a matter of careful bookkeeping, and we leave it as an exercise for the reader. $\square$

Behind the technical smokescreen, what the Implicit Function Theorem says is actually very natural: if at a point $\mathbf{x} \in \mathbb{R}^n$ there are m coordinates such that the determinant of the matrix of the corresponding partial derivatives is nonzero, then locally around $\mathbf{x}$ you may choose the complementary $n - m$ coordinates to be local coordinates for the level set of F through $\mathbf{x}$. The natural projection to these coordinates gives a local chart for the level set around $\mathbf{x}$. With this in mind, the following result should seem very natural.

Definition 2.5.2. Let $F : \mathbb{R}^n \to \mathbb{R}^m$ be a smooth function. A point $\mathbf{a} \in \mathbb{R}^m$ is called a **regular value** for F, if for every $\mathbf{x} \in \mathbb{R}^n$ such that $F(\mathbf{x}) = \mathbf{a}$, the differential $dF(\mathbf{x})$ is a surjective linear function.

Theorem 2.5.3. *Let $F : \mathbb{R}^n \to \mathbb{R}^m$ be a smooth function and $\mathbf{a} \in \mathbb{R}^m$ a regular value for F. Then $F^{-1}(\mathbf{a})$ is a smooth manifold.*

Proof   We begin by noting that $F^{-1}(\mathbf{a})$ is Hausdorff since it is a subset of $\mathbb{R}^n$ with the induced subset topology.

Since $\mathbf{a}$ is a regular value for F, for any $\mathbf{x}$ in the level set of $\mathbf{a}$, Theorem 2.5.1 applies: the pair $(U_{\mathbf{x}}, \pi_{n-m})$ gives a local chart for $F^{-1}(\mathbf{a})$ around $\mathbf{x}$.

Now assume $\mathbf{x}$ and $\mathbf{x}'$ are in the level set of $\mathbf{a}$ and $V_{\mathbf{x}} \cap V_{\mathbf{x}'} \cap F^{-1}(\mathbf{a}) \neq \varnothing$. Then the transition function $T_{\mathbf{x},\mathbf{x}'} = \varphi_{\mathbf{x}} \circ \varphi_{\mathbf{x}'}^{-1} = \pi_{n-m} \circ f_{\mathbf{x}'}$ (restricted to the appropriate domain of definition) is a composition of smooth functions and hence it is smooth.   □

Exercise 2.5.1. Consider the function $F : \mathbb{R}^3 \to \mathbb{R}$ defined by:

$$F(x, y, z) = \left(2 - \sqrt{x^2 + y^2}\right)^2 + z^2.$$

Show that 1 is a regular value for F and hence $F^{-1}(1)$ is a smooth manifold. It is in fact a familiar surface – try to recognize it.

3

Riemann Surfaces

In Chapter 1 we saw that studying maximal domains of definition for complex analytic functions naturally led us to look at geometric spaces which are locally indistinguishable from $\mathbb{C}$, but globally are different from $\mathbb{C}$. In Chapter 2 we saw that the notion of manifolds formalizes the idea of "spaces formed by gluing together Euclidean spaces by identifying open sets". We now revisit the complex spaces studied in Chapter 1 as a particular class of manifolds, called *Riemann Surfaces*. The name is given after the mathematician Bernhard Riemann (1826–66), who introduced this point of view.

Definition 3.0.4. A **Riemann Surface** is a complex analytic manifold of dimension 1.

In a few more words, "X is a Riemann Surface" means:

1. X is a Hausdorff, connected topological space;
2. For all $x \in X$ there is a homeomorphism $\varphi_x : U_x \to V_x$, where U_x is an open neighborhood of $x \in X$ and V_x is an open set in $\mathbb{C}$;
3. For any U_x, U_y such that $U_x \cap U_y \neq \varnothing$ the transition function

$$T_{y,x} := \varphi_y \circ \varphi_x^{-1} : \varphi_x(U_x \cap U_y) \to \varphi_y(U_x \cap U_y)$$

is holomorphic.

Remark 3.0.5. The choice of requiring connectedness in Definition 3.0.4 is made purely for exposition convenience. One can define a **disconnected Riemann Surface** simply by removing the connectedness assumption. When we work with disconnected Riemann Surfaces in this book we will always qualify them as such.

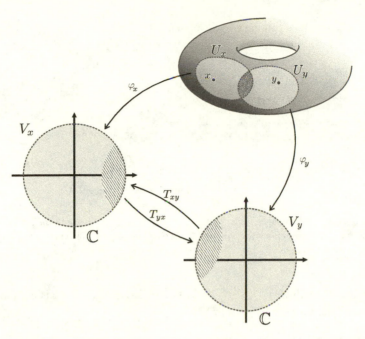

Figure 3.1 A Riemann Surface with two local charts and transition functions

We saw in Corollary 1.1.3 that a holomorphic function preserves orientation when thought of as a differentiable function from the (real) plane to itself. Since all transition functions are holomorphic, any atlas is a **positive atlas**. Therefore, topologically a Riemann Surface is an **orientable surface**.

The remainder of the chapter is dedicated to providing several examples of Riemann Surfaces. We begin by revisiting the examples seen in Chapter 1 and several simple examples coming from complex analysis. We then move to compact Riemann Surfaces, which are the heroes of this book.

3.1 Examples of Riemann Surfaces

3.1.1 The Riemann Surface of the Square Root

In Chapter 1, to create an honest domain for the function $z^{1/2}$ we altered the topology of two copies of $\mathbb{C} \smallsetminus 0$, that we denoted X^+ and X^-: the resulting (Hausdorff) topological space X can and should be thought of as having cut and re-glued the two punctured complex planes along the positive real numbers (see Example 1.4.3 for details). We show that X is a Riemann Surface by exhibiting an atlas with four local charts: we begin by defining the inverse functions to the charts.

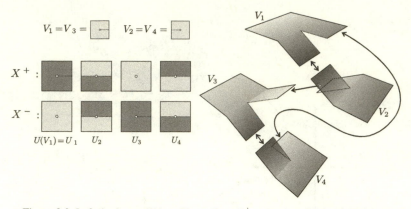

Figure 3.2 Left: images $\iota_i(V_i) = U_i \subset X = X^+ \sqcup X^- / \sim$; Right: the atlas on X provides instructions for "gluing" the sets V_i by identifying points using the transition functions

Define $V_1 = V_3 := \mathbb{C} \smallsetminus \mathbb{R}^{\geq 0}$, $V_2 = V_4 := \mathbb{C} \smallsetminus \mathbb{R}^{\leq 0}$, and consider the four injective functions $\iota_i : V_i \to X$:

$$\iota_1(z_1) = z_1 \in X^+,$$

$$\iota_2(z_2) = \begin{cases} z_2 \in X^+ & \text{if Im}\,(z_2) \leq 0 \\ z_2 \in X^- & \text{if Im}\,(z_2) > 0 \end{cases},$$

$$\iota_3(z_3) = z_3 \in X^-,$$

$$\iota_4(z_4) = \begin{cases} z_4 \in X^- & \text{if Im}\,(z_4) \leq 0 \\ z_4 \in X^+ & \text{if Im}\,(z_4) > 0 \end{cases}. \tag{3.1}$$

Call U_i the image $\iota_i(V_i) \subset X$, and $\varphi_i : U_i \to V_i$ the inverse function of ι_i. See Figure 3.2.

Exercise 3.1.1. Prove that the sets U_i are open sets in X (you must use the topology on X described in Example 1.4.3). Show that the collection $\{U_i\}_{i=1,\ldots,4}$ covers X, and that the functions φ_i are homeomorphisms.

It remains to show that the transition functions are holomorphic. We consider one transition function – all other checks are analogous.

The intersection $U_1 \cap U_2$ consists of all points in X^+ whose imaginary part is negative. The transition function $T_{21} : \varphi_1(U_1 \cap U_2) = \{z \in V_1 | \text{Im}\, z < 0\} \to \varphi_2(U_1 \cap U_2) = \{z \in V_2 | \text{Im}\, z < 0\}$ maps $z_1 \mapsto z_1$ and thus gives the change of coordinates $z_2 = z_1$, which is holomorphic. This completes the proof that X is a Riemann Surface.

Exercise 3.1.2. Exhibit the domains for $z^{\frac{1}{n}}$ and $\log z$ as Riemann Surfaces. Note that, as a set, $\log z = w \in \mathbb{C} | e^w = z$.

Remark 3.1.1. We pause to note an alternative technique for constructing a Riemann Surface. Namely, instead of assigning local charts to a topological space X, one can start with a collection $\{V_\alpha\}_{\alpha \in A}$ of open sets in $\mathbb{C}$ together with compatible transition functions $t_{\beta\alpha} : U_\alpha \to V_\beta$, where $U_\alpha \subseteq V_\alpha$ is an open subset and $t_{\beta\alpha}$ is a homeomorphism onto its image. The space X is then constructed by gluing the sets V_α using the transition functions. Formally, X is the identification space

$$X \cong \frac{\coprod_{\alpha \in A} V_\alpha}{\sim},$$

where $\sim$ is the equivalence relation generated by $x \sim t_{\beta\alpha}(x)$ for all $x \in V_\alpha$ and all $\alpha \in A$. See Miranda (1995, Section I.2) for more details.

3.1.2 Graphs of Complex Functions $f(z)$

A class of examples of Riemann Surfaces is given by graphs of continuous complex functions. Let $f(z)$ be a continuous function mapping $\mathbb{C}$ to $\mathbb{C}$. The graph of f is the set $\Gamma_f := \{(z, f(z)) | z \in \mathbb{C}\} \subset \mathbb{C} \times \mathbb{C}$ given the subspace topology. In Figure 3.3 we visualize all the relevant maps.

First we note that Γ_f is Hausdorff since $\mathbb{C} \times \mathbb{C}$ is. The graph of f is naturally given the structure of a Riemann Surface by an atlas with one chart, namely all of Γ_f; the local coordinate function is the first projection map $\varphi := \pi_1|_{\Gamma_f}$, which sends $(z, f(z))$ to z. To define a Riemann Surface structure, φ must be

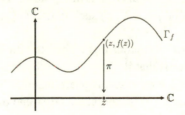

Figure 3.3 Schematic picture of a graph Γ_f

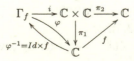
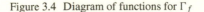

Figure 3.4 Diagram of functions for Γ_f

a homeomorphism onto its image (which is all of $\mathbb{C}$). The map $\varphi = \pi \circ i$ is the composition of i, the natural inclusion of Γ_f into $\mathbb{C} \times \mathbb{C}$, and π_1, the projection onto the first factor. Both i and π_1 are continuous maps, and thus so is φ.

To show that φ^{-1} is continuous, we observe that $\varphi^{-1} = Id \times f : \mathbb{C} \to \mathbb{C} \times \mathbb{C}$. Since the identity function and f are both continuous functions, it follows that their product is continuous.

Given that there is only one chart, the holomorphicity of transition functions is trivially satisfied. Thus Γ_f is a Riemann Surface. Once we discuss maps (and, in particular, isomorphisms) of Riemann Surfaces, we will see that all graphs are isomorphic to the Riemann Surface $X = \mathbb{C}$ (Exercise 4.1.7).

3.1.3 Algebraic Curves

The complex analysis version of the Implicit Function Theorem (Theorem 2.5.1) implies that the inverse image of a regular value for an analytic function is a complex manifold. In particular, if $f : \mathbb{C}^{n+1} \to \mathbb{C}^n$ is a holomorphic function such that $0 \in \mathbb{C}^n$ is a regular value for f, then $f^{-1}(0)$ is a complex analytic manifold of dimension 1, i.e. a Riemann Surface.

When the holomorphic function $f = (f_1, \ldots, f_n)$ is given by a collection of n polynomials in $n + 1$ variables, the Riemann Surfaces arising as inverse images of regular values are also called **affine algebraic curves**. We study in detail the plane curve case, i.e. when $n = 1$. For an analogy, recall that over the real numbers, the set $\{x^2 + y^2 - 1 = 0\}$ is the unit circle: it lives in $\mathbb{R}^2$ and is a manifold of real dimension 1.

Definition 3.1.2. For any $p(x, y) \in \mathbb{C}[x, y]$, the set $V(p) := \{(x, y) | p(x, y) = 0\} \subset \mathbb{C}^2$ is called an **affine plane curve**. We say that $V(p)$ is **smooth** if there is no $(x_0, y_0) \in V(p)$ such that $\frac{\partial p}{\partial x}(x_0, y_0) = 0 = \frac{\partial p}{\partial y}(x_0, y_0)$.

We now give a sketch of a proof that a smooth affine plane curve is a Riemann Surface. The idea is that if $V(p)$ is smooth, then locally it can be seen as a graph, and these local expressions patch together well. To be precise, let $(x_0, y_0) \in V(p)$. Since $V(p)$ is smooth, at least one of $\frac{\partial p}{\partial x}, \frac{\partial p}{\partial y}$ is nonzero at (x_0, y_0). Say that $\frac{\partial p}{\partial y}(x_0, y_0) \neq 0$. By the Implicit Function Theorem there is a neighborhood $U_{(x_0, y_0)} \subset \mathbb{C}^2$, a neighborhood $V_{x_0} \subset \mathbb{C}$, and a holomorphic function $f(x) : V_{x_0} \to \mathbb{C}$ such that $V(p) \cap U_{(x_0, y_0)} = \{(x, f(x)) | x \in V_{x_0}\}$, the graph of f (see Figure 3.5).

We get a local chart on $V(p)$ around (x_0, y_0) as in Section 3.1.2, by setting $\varphi_{(x_0, y_0)} : V(p) \cap U_{(x_0, y_0)} \to V_{x_0}$ to be projection to the first factor: $\varphi_{(x_0, y_0)}(x, f(x)) = x$.

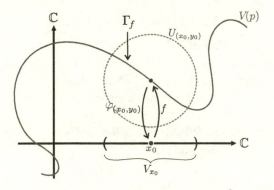

Figure 3.5 Local chart around $(x_0, y_0) \in V(p)$ where $\frac{\partial p}{\partial y}(x_0, y_0) \neq 0$

Finally, we show that transition functions are holomorphic. Assume $U_{(x_0,y_0)} \cap U_{(x_1,y_1)} \cap V(p) \neq \varnothing$. If $\varphi_{(x_0,y_0)}$ and $\varphi_{(x_1,y_1)}$ are both projections to the same axis, the transition function $\varphi_{(x_1,y_1)} \circ \varphi_{(x_0,y_0)}^{-1}$ is just the identity function restricted to the appropriate domain in $\mathbb{C}$. Assume now that $\varphi_{(x_0,y_0)}$ is projection onto the x-axis and that $\varphi_{(x_1,y_1)}$ is projection onto the y-axis. Then the set $U_{(x_0,y_0)} \cap U_{(x_1,y_1)} \cap V(p)$ is simultaneously on the graph of a holomorphic function $f_0(x)$ and of a holomorphic function $f_1(y)$. The transition functions are then $\varphi_{(x_1,y_1)} \circ \varphi_{(x_0,y_0)}^{-1} = f_0(x)$ and $\varphi_{(x_0,y_0)} \circ \varphi_{(x_1,y_1)}^{-1} = f_1(y)$ restricted to appropriate domains, which are holomorphic.

Exercise 3.1.3. Spell out the details of the above argument that a smooth affine plane curve is a Riemann Surface.

3.2 Compact Riemann Surfaces

We turn our attention to compact Riemann Surfaces, which are the heroes of our story. Compactness is a strong constraint on the geometry of surfaces, and it is responsible for some of the rich structure of the theory of analytic functions among Riemann Surfaces.

From a topological point of view, a compact Riemann Surface X is a compact orientable surface. By the classification of compact surfaces theorem (Theorem 2.4.3), it is homeomorphic to the connected sum of g tori. The integer g is called the **genus** of X: it can be thought of as the number of "handles" that are attached to a sphere to obtain X.

We look at some examples of compact Riemann Surfaces, starting from the Complex Projective Line, or Riemann Sphere.

3.2.1 Complex Projective Line

In Chapter 2 we introduced $\mathbb{P}^1(\mathbb{R})$ as a manifold whose points parameterize lines through the origin in $\mathbb{R}^2$. Similarly, $\mathbb{P}^1(\mathbb{C})$ is a manifold whose points parameterize complex one-dimensional linear subspaces of $\mathbb{C}^2$. The construction is analogous to Section 2.3; here we present $\mathbb{P}^1(\mathbb{C})$ as a Riemann Surface by gluing together two copies of $\mathbb{C}$.

Let $U_1 = U_2 := \mathbb{C}$ and define $g : U_1 \smallsetminus \{0\} \to U_2 \smallsetminus \{0\}$ by $g(z) = 1/z$. We define $\mathbb{P}^1(\mathbb{C})$ as the identification space

$$\mathbb{P}^1(\mathbb{C}) := U_1 \bigcup_g U_2 = \frac{U_1 \coprod U_2}{z \sim g(z)}.$$

The idea is to have two copies of $\mathbb{C}$ standing side by side. Then, holding both 0s still, fold the copies towards each other, identifying each nonzero point to the point on the other copy corresponding to its inverse.

Exercise 3.2.1. Show that, as a set, $\mathbb{P}^1(\mathbb{C})$ is $\mathbb{C}$ plus a point. Prove that $\mathbb{P}^1(\mathbb{C})$ is a Hausdorff topological space. Prove that $\mathbb{P}^1(\mathbb{C})$ is the one-point compactification of $\mathbb{C}$, and therefore homeomorphic to a sphere. In complex analysis, it is called the **Riemann Sphere**.

For $i = 1, 2$, we denote by $[U_i]$ the image of the set U_i after the identification by g: note that $[U_i]$ is an open set in $\mathbb{P}^1(\mathbb{C})$. Define the local coordinate functions $\varphi_i : [U_i] \to U_i$ by $\varphi_i(p) = z_i$, where z_i is the complex number in U_i such that $[z_i] = p$. Both φ_1 and φ_2 are homeomorphisms.

We now consider transition functions; specifically, we consider T_{21}. The intersection $[U_1] \cap [U_2] = [U_1 \smallsetminus \{0\}] = [U_2 \smallsetminus \{0\}]$ and its image through φ_1 is $\varphi_1([U_1] \cap [U_2]) = \mathbb{C} \smallsetminus \{0\}$. This is the domain of $T_{21} = \varphi_2 \circ \varphi_1^{-1}$, and for $z_1 \neq 0$ we have

$$z_1 \overset{\varphi_1^{-1}}{\longmapsto} [z_1] = [z_2 = g(z_1) = 1/z_1] \overset{\varphi_2}{\longmapsto} z_2 = 1/z_1.$$

Since T_{21} has a pole only at $z_1 = 0$, it is holomorphic on $\mathbb{C} \smallsetminus \{0\}$. A symmetric computation shows that T_{12} is holomorphic, and we have that $\mathbb{P}^1(\mathbb{C})$ is a Riemann Surface. Since $\mathbb{P}^1(\mathbb{C})$ is homeomorphic to a sphere, its genus is 0.

Exercise 3.2.2. Here is an alternative way of showing that $\mathbb{P}^1(\mathbb{C})$ is a compact space. Consider a three-dimensional real sphere $S^3 \subset \mathbb{C}^2$ as the locus of points that are of distance 1 from the origin. Given any point $p \in S^3$, there is a unique complex line ℓ_p through the origin and p. We therefore get a function:

$$H: \quad S^3 \quad \rightarrow \quad \mathbb{P}^1(\mathbb{C})$$
$$p \quad \mapsto \quad \ell_p.$$

Check that H is a continuous and surjective function. Since S^3 is compact (closed and bounded) and the continuous image of a compact set is compact, this proves that $\mathbb{P}^1(\mathbb{C})$ is compact.

Check that for any point $\ell_p \in \mathbb{P}^1(\mathbb{C})$, the inverse image $H^{-1}(\ell_p)$ is a circle. The map H, which realizes the three-dimensional sphere as a circle fibration over a two-dimensional sphere, is famously known as the *Hopf fibration*.

Remark 3.2.1. In this section we saw that $\mathbb{P}^1(\mathbb{C})$ is an example of a Riemann Surface of genus 0. In fact, it is the only example. This follows from an important theorem in algebraic geometry, the Riemann–Roch theorem (Miranda, 1995, page 185), which guarantees that a genus 0 compact Riemann Surface admits a degree 1 meromorphic function. One then verifies that such a function defines an isomorphism with $\mathbb{P}^1(\mathbb{C})$.

Exercise 3.2.3. One can take U_1 and U_2 as before and glue them together using $\tilde{g} : U_1 \smallsetminus \{0\} \rightarrow U_2 \smallsetminus \{0\}$ where $\tilde{g}(z) = z$. Show that the resulting topological space, $X = U_1 \cup_{\tilde{g}} U_2$, is not Hausdorff (and thus is *not* a Riemann Surface!). It is often called the **complex plane with doubled origin**.

3.2.2 Complex Tori

Complex tori are examples of compact Riemann Surfaces of genus 1. In fact, any compact Riemann Surface of genus 1 is isomorphic to a complex torus (see Silverman and Tate (1992)).

Definition 3.2.2. Let τ_1 and τ_2 be two complex numbers which are linearly independent over $\mathbb{R}$ (i.e. they don't lie on the same real line through 0 in $\mathbb{C}$). The set of all integral linear combinations of τ_1 and τ_2

$$\Lambda = \{n\tau_1 + m\tau_2 \mid n, m \in \mathbb{Z}\} \subset \mathbb{C}$$

is called a **lattice** of complex numbers.

We will see in Exercise 4.5.5 that we may assume that $\tau_1 = 1$ and $\text{Im}(\tau_2) > 0$, so we make the simplifying assumption that a lattice has the form $\Lambda = \{n + m\tau \mid n, m \in \mathbb{Z}, \tau \in \mathbb{H}\} \subset \mathbb{C}$. Here $\mathbb{H}$ is the upper half-plane $\mathbb{H} = \{z \in \mathbb{C} \mid \text{Im}(z) > 0\}$.

Consider the quotient space $T = \mathbb{C}/\Lambda$, i.e. the identification space $T = \mathbb{C}/\sim$ where $z_1 \sim z_2$ if and only if $z_2 = z_1 + w$ for some $w \in \Lambda$. The natural

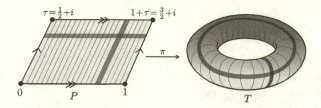

Figure 3.6 Torus together with identification polygon and loops

projection map $\pi : \mathbb{C} \to T$ given by $\pi(z) = [z]$ induces a natural quotient topology on T, i.e. $V \subset T$ is open in T if and only if $\pi^{-1}(V)$ is open in $\mathbb{C}$.

Exercise 3.2.4. Denote by P the closed parallelogram with vertices 0, 1, τ, $1+\tau$. Show that for any $z \in \mathbb{C}$ there is a $z' \in P$ with $z \sim z'$. This shows that $\pi|_P : P \to T$ is onto, and hence we can restrict our attention to P in order to understand the geometry of T.

By considering the residual identification of points in P (defined in Exercise 3.2.4) we see that T is topologically a torus. If z is in P but is not on the boundary, then z is not equivalent to any other point of P. If z is on the line segment through 0 and τ, then $z \sim z + 1 \in P$. Also, if z is on the line segment through 0 and 1, then $z \sim z + \tau \in P$.

The topology of T is described by the identification polygon in Figure 3.6.

Exercise 3.2.5. Prove that π is an open map, i.e. that V open in $\mathbb{C}$ implies that $\pi(V)$ is open in T.

We now endow T with a complex structure. Exercise 3.2.5 shows that if π restricted to a subset $V \subset \mathbb{C}$ is one-to-one, then it is a homeomorphism onto its image in T. In this case, $(\pi|_V)^{-1}$ is also a homeomorphism from the image of $\pi|_V$ to V, and we may use $(\pi|_V)^{-1}$ as a chart of T.

Exercise 3.2.6. Find a real number r (depending on τ) such that, for any $z \in \mathbb{C}$, π restricted to $B_r(z)$, a ball of radius r centered at z, is a one-to-one map.

Given r as in Exercise 3.2.6 and $z \in \mathbb{C}$, define

$$U_z := \pi(B_r(z)) \subset T \quad \text{and} \quad \varphi_z := (\pi|_{B_r(z)})^{-1}.$$

We claim that the collection:

$$\mathcal{A} = \{(U_z, \varphi_z) \,|\, z \in \mathbb{C}\}$$

forms an atlas for T. Now $\mathcal{A}$ certainly gives a (highly redundant) cover of T by open sets, and we have arranged for the maps φ_z to be homeomorphisms onto their images. Assume that $U_{z_1} \cap U_{z_2} \neq \varnothing$; for $i = 1, 2$, denote by (α_i, β_i) the unique pair of real numbers such that $z_i = \alpha_i + \beta_i \tau$. We have that

$$T_{21}(z) = \varphi_{z_2} \circ \varphi_{z_1}^{-1}(z) = z + k,$$

where $k = (\lfloor \alpha_2 \rfloor - \lfloor \alpha_1 \rfloor) + (\lfloor \beta_2 \rfloor - \lfloor \beta_1 \rfloor) \tau$ is just a constant depending on z_1 and z_2. Therefore the transition function T_{21} is holomorphic. This proves that $\mathcal{A}$ is an atlas and therefore T is a Riemann Surface.

3.2.3 Projective Curves

In section 3.1.3 we constructed Riemann Surfaces in the complex plane as level sets for a regular value of a polynomial function. In a similar fashion we wish to construct compact Riemann Surfaces as closed subsets of the complex projective plane $\mathbb{P}^2(\mathbb{C})$. The first obstacle arises from the fact that polynomials in the homogeneous coordinates of the projective plane do not define functions on $\mathbb{P}^2(\mathbb{C})$.

Example 3.2.3. Consider the polynomial $p(x, y, z) = x^2 + y + z + 1$. Note that

$$p(1, 1, 1) = 4 \neq 7 = p(2, 2, 2).$$

Since $[1 : 1 : 1] = [2 : 2 : 2]$ are the same point in $\mathbb{P}^2(\mathbb{C})$, p is attempting to assign two different outputs to the same input, violating the definition of a function.

Geometrically, the coordinates of any point belonging to a complex line ℓ through the origin in $\mathbb{C}^3$ can be chosen to represent the point corresponding to ℓ in $\mathbb{P}^2(\mathbb{C})$; hence, for a polynomial in three variables, to give a well-defined function on $\mathbb{P}^2(\mathbb{C})$ one would need the polynomial to remain constant along lines through the origin. Alas, the only polynomials that satisfy such a condition are globally constant polynomials, which don't make for particularly exciting functions...

Recall, however, that in defining an affine plane curve we only care about points where the polynomial vanishes. We therefore now seek polynomials in three variables that vanish along lines through the origin in $\mathbb{C}^3$. Luckily there is a large collection of such polynomials, giving us a rich set of examples of compact Riemann Surfaces.

Definition 3.2.4. A polynomial $P \in \mathbb{C}[X, Y, Z]$ is said to be **homogeneous of degree** d if any of the following equivalent conditions is satisfied:

1. Every monomial of P has degree d;
2. For every $t \in \mathbb{C}$,

$$P(tX, tY, tZ) = t^d P(X, Y, Z);$$

3.

$$X \frac{\partial P}{\partial X} + Y \frac{\partial P}{\partial Y} + Z \frac{\partial P}{\partial Z} = dP$$

(here dP denotes the number d times the polynomial P).

Exercise 3.2.7. Show that conditions 1, 2 and 3 in Definition 3.2.4 are indeed equivalent. Condition 3 is sometimes called **Euler's Identity**.

Exercise 3.2.8. Show that if $P \in \mathbb{C}[X, Y, Z]$ is a homogeneous polynomial, then the set of points $[X : Y : Z] \in \mathbb{P}^2(\mathbb{C})$ where P vanishes is well defined. We call this set the **vanishing locus** of P.

Definition 3.2.5. Given $P \in \mathbb{C}[X, Y, Z]$, a homogeneous polynomial of degree d, the set

$$V(P) := \{[X : Y : Z] \in \mathbb{P}^2(\mathbb{C}) | P(X, Y, Z) = 0\}$$

is called a **plane projective curve** of degree d. If

$$\left\{ (X, Y, Z) \in \mathbb{C}^3 \,\middle|\, \frac{\partial P}{\partial X} = \frac{\partial P}{\partial Y} = \frac{\partial P}{\partial Z} = 0 \right\} \subseteq \{(0, 0, 0)\}$$

then $V(P)$ is said to be **smooth**.

Refer back to Figure 2.3 on page 19 to visualize what is going on. The bottom of the figure consists of a cone in three-dimensional space and a slice of the cone on a canvas. Notice that the cone is a two-dimensional object, but the curve on the canvas is one-dimensional. The vanishing locus of P in $\mathbb{C}^3$, the set $\{(x, y, z) \in \mathbb{C}^3 | P(x, y, z) = 0\}$, is a cone, i.e. consists of (complex) lines through the origin which together form an object of complex dimension 2. In the vanishing locus of P in $\mathbb{P}^2(\mathbb{C})$, i.e. $V(P)$, each line in the cone represents one point in $\mathbb{P}^2(\mathbb{C})$ and the slice on the canvas represents the image of $V(P)$ through a local chart of $\mathbb{P}^2(\mathbb{C})$. The image has now complex dimension one, and is hence called a curve.

Proposition 3.2.6. *A smooth projective plane curve $V(P)$ is a compact Riemann Surface.*

Proof We first show that $V(P)$ is compact by showing that $V(P)$ is a closed set in $\mathbb{P}^2(\mathbb{C})$, which is a compact topological space. Consider the diagram

$$
\begin{array}{ccc}
\mathbb{C}^3 \smallsetminus \{(0,0,0)\} & \xrightarrow{\ P\ } & \mathbb{C}\,, \\
\downarrow{\scriptstyle \pi} & & \\
\mathbb{P}^2(\mathbb{C}) & &
\end{array}
$$

where π is the natural projection function and P is the (continuous) function defined by the homogeneous polynomial P (i.e. $P : (X, Y, Z) \mapsto P(X, Y, Z)$). By definition, $V(P)$ is a closed subset of $\mathbb{P}^2(\mathbb{C})$ if $\pi^{-1}(V(P))$ is closed in $\mathbb{C}^3 \smallsetminus (0,0,0)$. But

$$
\pi^{-1}(V(P)) = P^{-1}(0)
$$

is the inverse image of the closed set $\{0\} \subset \mathbb{C}$, therefore it is closed.

To prove that $V(P)$ is a Riemann Surface, it is sufficient to show that its intersection with any of the coordinate open sets of $\mathbb{P}^2(\mathbb{C})$ is a Riemann Surface. Consider (without loss of generality) the chart

$$
U_Z = \{[X : Y : Z] | Z \neq 0\} \subseteq \mathbb{P}^2(\mathbb{C})
$$

with affine coordinates

$$
(x, y) = \varphi_Z(X, Y, Z) = (X/Z, Y/Z).
$$

The set $\varphi_Z(V(P) \cap U_Z)$ is equal to $V(p)$, where $p(x, y) := P(x, y, 1)$ is called the dehomogenization of P with respect to Z. For any $(x, y) \in \mathbb{C}^2$

$$
\frac{\partial p}{\partial x}(x, y) = \frac{\partial P}{\partial X}(x, y, 1) \tag{3.2}
$$

$$
\frac{\partial p}{\partial y}(x, y) = \frac{\partial P}{\partial Y}(x, y, 1). \tag{3.3}
$$

We claim that there can be no $(\tilde{x}, \tilde{y}) \in \mathbb{C}^2$ such that

$$
p(\tilde{x}, \tilde{y}) = \frac{\partial p}{\partial x}(\tilde{x}, \tilde{y}) = \frac{\partial p}{\partial y}(\tilde{x}, \tilde{y}) = 0. \tag{3.4}
$$

The claim implies that $V(p)$ is a smooth affine plane curve and therefore a Riemann Surface as in Section 3.1.3. Since the restriction of $V(P)$ with any affine chart is a Riemann Surface, so is $V(P)$.

To prove the claim, assume there is $(\tilde{x}, \tilde{y}) \in \mathbb{C}^2$ satisfying the system of equations in (3.4). By (3.2), (3.3), and the smoothness of $V(P)$, it must be that

$$\frac{\partial P}{\partial Z}(\tilde{x}, \tilde{y}, 1) \neq 0.$$

But now Euler's Identity gives us a contradiction, since

$$0 \neq \frac{\partial P}{\partial X}(\tilde{x}, \tilde{y}, 1) + \frac{\partial P}{\partial Y}(\tilde{x}, \tilde{y}, 1) + \frac{\partial P}{\partial Z}(\tilde{x}, \tilde{y}, 1) = dP(\tilde{x}, \tilde{y}, 1) = 0.$$

$\square$

Example 3.2.7 (Conics). Consider the homogeneous polynomial

$$P(X, Y, Z) = X^2 + Y^2 - Z^2.$$

We check that $V(P)$ is a smooth projective plane curve of degree 2, also known as a **conic**. Indeed,

$$\frac{\partial P}{\partial X} = 2X, \quad \frac{\partial P}{\partial Y} = 2Y, \quad \frac{\partial P}{\partial Z} = -2Z$$

and $(0, 0, 0)$ is the only point where all partial derivatives vanish simultaneously.

We observe that if we dehomogenize P with respect to Z (by defining $x = X/Z$, $y = Y/Z$), we recognize the equation of a circle, whereas if we dehomogenize with respect to X or Y we obtain the equation of a hyperbola. This has to do with the fact that affine plane conics are obtained as plane sections of a cone: before we identify all the points on a line through the origin, the solutions of $P(X, Y, Z) = X^2 + Y^2 - Z^2 = 0$ in $\mathbb{C}^3$ give a cone, and slicing it with different planes just amounts to restricting the projective curve $V(P)$ to different affine charts. (Imagine holding the cone at the bottom of Figure 2.3 steady, then moving the canvas to obtain different slices of the cone.) We will see in Exercise 4.1.8 that all smooth conics are isomorphic to $\mathbb{P}^1(\mathbb{C})$, and are therefore Riemann Surfaces of genus 0.

Exercise 3.2.9. Consider an arbitrary homogeneous polynomial $P \in \mathbb{C}[X, Y, Z]$ of degree 2. What are the conditions on the coefficients of P for $V(P)$ to be a smooth conic? When these conditions are not met, $V(P)$ is called a **degenerate conic**. Show that there are two distinct types of degenerate conic.

Exercise 3.2.10. Recall from linear algebra that a homogeneous polynomial $P \in \mathbb{C}[X, Y, Z]$ of degree 2 naturally defines a quadratic form which can be represented in a unique way by a 3×3 symmetric matrix A_P. What linear algebraic conditions on A_P correspond to $V(P)$ being a smooth conic or each of the two types of degenerate conics?

Exercise 3.2.11. A **projectivity** is a map $\varphi : \mathbb{P}^2(\mathbb{C}) \to \mathbb{P}^2(\mathbb{C})$ defined by $\varphi([X : Y : Z]) = [p_1 : p_2 : p_3]$, where the p_is are three $\mathbb{C}$-linearly independent linear homogeneous polynomials in X, Y, Z (it can be thought of as a "change of coordinates" in the projective plane). Two subsets of $\mathbb{P}^2(\mathbb{C})$ are called **projectively equivalent** if there exists a projectivity that sends one to the other. Show that any two smooth complex conics are projectively equivalent.

Example 3.2.8 (Elliptic Curves). Consider a polynomial P of the form:

$$P(X, Y, Z) = Y^2 Z - (X - \alpha_1 Z)(X - \alpha_2 Z)(X - \alpha_3 Z),$$

where $\alpha_1, \alpha_2, \alpha_3$ are three distinct complex numbers. Note that the partial derivative with respect to Y is $\partial P / \partial Y = 2YZ$, which is zero only if $Z = 0$ or $Y = 0$. We show that $V(P)$ is a smooth projective curve by considering the cases $Z = 0$, $Y = 0$ and finding in each case a non-vanishing partial derivative.

Case 1: If $Z = 0$, then the only point in $\mathbb{P}^2(\mathbb{C})$ belonging to $V(P)$ is $[0 : 1 : 0]$. But we have

$$\frac{\partial P}{\partial Z} = Y^2 + Q(X, Z),$$

where $Q(X, Z)$ is a homogeneous degree 2 polynomial in X and Z – and so $\partial P / \partial Z(0, 1, 0) = 1 \neq 0$.

Case 2: If $Y = 0$, then the points belonging to $V(P)$ are $[\alpha_1 : 0 : 1]$, $[\alpha_2 : 0 : 1]$, and $[\alpha_3 : 0 : 1]$. For $i = 1, 2, 3$

$$\frac{\partial P}{\partial X}(\alpha_i, 0, 1) \neq 0$$

follows from the facts that the α_i are distinct. This proves that $V(P)$ is a smooth projective curve of degree 3, also known as an **elliptic curve**.

Elliptic curves are Riemann Surfaces of genus 1. The connection between elliptic curves and complex tori is a very beautiful and classical story which can be found, for example, in Silverman and Tate (1992).

One can generalize the notion of $V(P)$ to that of a **projective algebraic variety**: a subset of $\mathbb{P}^n(\mathbb{C})$ defined by the simultaneous vanishing of a collection of homogeneous polynomials. A projective algebraic variety of complex dimension 1 is called a **projective curve**, and when it is smooth it is a compact Riemann Surface.

Generically, one needs $n-1$ homogeneous polynomials in $n+1$ variables to cut out a one-dimensional set in $\mathbb{P}^n(\mathbb{C})$ (we make this precise in Exercise 3.2.12); however, there exist projective curves that cannot be "cut out" by the "right" number of equations; we see the first example in Exercise 3.2.13.

Exercise 3.2.12. Let $P_1, \ldots, P_{n-1}$ be homogeneous polynomials in $n+1$ variables, and define

$$V(\mathbf{P}) = \{[X_0 : \ldots : X_n] | P_1(X_0, \ldots, X_n) = \cdots = P_{n-1}(X_0, \ldots, X_n) = 0\} \subset \mathbb{P}^n(\mathbb{C}).$$

Show that if, for every point $\mathbf{X} \in V(\mathbf{P})$, the matrix of partial derivatives

$$\begin{bmatrix} \frac{\partial P_1}{\partial X_0}(\mathbf{X}) & \cdots & \frac{\partial P_1}{\partial X_n}(\mathbf{X}) \\ \vdots & \cdots & \vdots \\ \frac{\partial P_{n-1}}{\partial X_0}(\mathbf{X}) & \cdots & \frac{\partial P_{n-1}}{\partial X_n}(\mathbf{X}) \end{bmatrix}$$

has maximal rank $n-1$, then by the Implicit Function Theorem $V(P)$ is a compact Riemann Surface.

Exercise 3.2.13. Consider the function

$$\varphi : \mathbb{P}^1(\mathbb{C}) \to \mathbb{P}^3(\mathbb{C})$$

defined in homogeneous coordinates by

$$\varphi([S : T]) = [S^3 : S^2T : ST^2 : T^3].$$

You may check, or just believe for now, that φ is a map of complex manifolds and that the image of φ is a Riemann Surface isomorphic to $\mathbb{P}^1(\mathbb{C})$. We call the image of φ the **twisted cubic** in $\mathbb{P}^3(\mathbb{C})$.

Denoting $[X : Y : Z : W]$ the homogeneous coordinates of $\mathbb{P}^3(\mathbb{C})$, consider the polynomials

$$P_1 = XW - YZ \tag{3.5}$$

$$P_2 = XZ - Y^2 \tag{3.6}$$

$$P_3 = YW - Z^2. \tag{3.7}$$

Show that the vanishing locus of P_1, P_2 and P_3 is precisely the twisted cubic, but that the vanishing locus of any two of the three polynomials is strictly larger. The twisted cubic is the first example of a projective curve which is not a **complete intersection**, i.e. which is not cut out by the expected number of equations.

4

Maps of Riemann Surfaces

In this chapter we develop and study the notion of functions between Riemann Surfaces. Because Riemann Surfaces are locally identified with $\mathbb{C}$, it is natural to require maps of Riemann Surfaces locally to be identified with holomorphic maps from $\mathbb{C}$ to $\mathbb{C}$. Such an identification, called a *local expression*, requires charts to be chosen both in the source and target space, and hence it is highly non-unique. While this might seem like a bug, it is actually a feature of the theory: at any point of the source, one may choose charts in such a way that the local expression is a power function. Philosophically, this means that the complex analytic notion of order of vanishing of a function at a point carries over to functions of Riemann Surfaces. Because "vanishing" is now an ill-defined notion (by changing coordinate chart in the target space, one may arrange the local expression of the function to take any value one wants at any given point), the name is changed to *ramification index*. In the case of a map of compact Riemann Surfaces, there is a finite set of points where the ramification index is strictly greater than 1, which allows a lot of information about a function of Riemann Surfaces to be extracted from its local behavior. For example, a map of Riemann Surfaces has generically a constant number of inverse images: this number is then called the *degree* of the map. Perhaps the high-point of this chapter is the *Riemann–Hurwitz* formula, which expresses a relation between the genera, the ramification and the degree of a map of Riemann Surfaces: in other words, these discrete invariants may not be chosen freely in order for a function to exist.

We conclude the chapter by studying two interesting classes of examples: maps from the projective line to itself, and maps of elliptic curves/complex tori.

4.1 Holomorphic Maps of Riemann Surfaces

Definition 4.1.1. Let X, Y be Riemann Surfaces and $f : X \to Y$ a set function.

1. We say that f is **holomorphic at** $x \in X$ if for every choice of charts $\varphi_x, \varphi_{f(x)}$ the function $\varphi_{f(x)} \circ f \circ \varphi_x^{-1}$ is holomorphic at x.

2. If $U \subset X$ is open, we say that f is **holomorphic on** U if f is holomorphic at each $x \in U$.

3. If f is holomorphic on $U = X$ we say that f is a **holomorphic map**.

The function $F = \varphi_{f(x)} \circ f \circ \varphi_x^{-1}$ is called a **local expression** for f (see Figure 4.1). From Definition 4.1.1 it would seem that in order to show that a function f is a holomorphic map, one must check the local expression for all possible combinations of local charts. However, Exercise 4.1.1 shows that it suffices to find one local expression that works.

Exercise 4.1.1. Show that a map of Riemann Surfaces $f : X \to Y$ is holomorphic at $x \in X$ if and only if there is *a* choice of charts $\varphi_x, \varphi_{f(x)}$ such that $\varphi_{f(x)} \circ f \circ \varphi_x^{-1}$ is holomorphic at x.

We begin with some trivial examples that we pose as exercises.

Exercise 4.1.2. Let X, Y be Riemann Surfaces and choose a point $y_0 \in Y$. Define the constant map $c : X \to Y$ by $c(x) = y_0$ for all $x \in X$. Show that c is a holomorphic map.

Exercise 4.1.3. Let X be a Riemann Surface. Define the **identity map** on X as the function $I_X : X \to X$ such that $I_X(x) = x$ for all $x \in X$. Prove that I_X is a holomorphic map.

Now for a more interesting example of a holomorphic function.

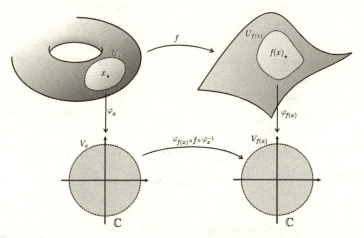

Figure 4.1 Schematic picture of a map of Riemann Surfaces with local expression

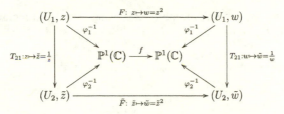

Figure 4.2 We check that the map f is holomorphic by checking that two local expressions, corresponding to charts that cover $\mathbb{P}^1(\mathbb{C})$, are holomorphic. Note that in our diagram we depict the inverses to the local coordinate functions, since the coordinate functions are only defined on open subsets of $\mathbb{P}^1(\mathbb{C})$

Example 4.1.2. Refer to Figure 4.2 to keep track of how all ingredients of this proof fit together. Recall from Exercise 3.2.1 that, as a set, $\mathbb{P}^1(\mathbb{C}) = \mathbb{C}$ plus a point. Identifying $\mathbb{C}$ with the image of the first affine chart $\varphi_1([U_1])$, the additional point corresponds to the image of 0 in the second affine chart U_2. We denote this point by ∞ and thus have $\mathbb{P}^1(\mathbb{C}) = \mathbb{C} \cup \{\infty\}$. Using this identification, define the function $f : \mathbb{P}^1(\mathbb{C}) \to \mathbb{P}^1(\mathbb{C})$ by $z \mapsto w = z^2$ and $\infty \mapsto \infty$. We show that f is a holomorphic map.

The way we described the set function f is by giving a local expression for it, using the chart U_1 for both the source and target $\mathbb{P}^1(\mathbb{C})$. We denoted by z the corresponding local coordinate on the source and by w the local coordinate on the target to avoid confusion. Then, since $w = F(z) = z^2$ is a holomorphic function on all of $\mathbb{C}$, f is holomorphic on the image of U_1.

All that is left to consider is whether f is holomorphic at ∞. We consider the local expression for f using the charts U_2 whose image contains ∞. We denote by $\tilde{z} = 1/z$ the corresponding local coordinate for the source, and $\tilde{w} = 1/w$ the coordinate on the target.

The local expression $\tilde{F}$ for f in these coordinates is obtained on the intersection of the charts by composing $F(z)$ with the transition functions for the local coordinates:

$$\tilde{F}(\tilde{z}) = \tilde{w} = \frac{1}{w} = \frac{1}{z^2} = \tilde{z}^2.$$

Since the point ∞ corresponds to $\tilde{z} = \tilde{w} = 0$, and we have $f(\infty) = \infty$ and $\tilde{F}(0) = (0)$, the local expression $\tilde{F}$ extends on the whole chart, and is in particular a holomorphic function at the point 0. This means that f is holomorphic at ∞ and completes the proof that f is a holomorphic function on all of $\mathbb{P}^1(\mathbb{C})$.

Exercise 4.1.4. Choose $a, b, c \in \mathbb{C}$ and consider the polynomial $p(z) = (z - a)(z - b)(z - c)$. Prove that the function $f : \mathbb{P}^1(\mathbb{C}) \to \mathbb{P}^1(\mathbb{C})$ given by $z \mapsto$

$p(z)$ and $\infty \mapsto \infty$ is a holomorphic map, where we again identify $\mathbb{P}^1(\mathbb{C}) = \mathbb{C} \cup \{\infty\}$ as in Example 4.1.2.

Exercise 4.1.5. Consider an arbitrary rational function

$$f(z) = \frac{p(z)}{q(z)},$$

for $p(z), q(z) \in \mathbb{C}[z]$, two polynomials with distinct roots. You may extend it to a function from $\mathbb{P}^1(\mathbb{C})$ to $\mathbb{P}^1(\mathbb{C})$ by defining:

- $f(\alpha) = \infty$, for α any root of $q(z)$;
- $f(\infty) = \lim_{z \to \infty} p(z)/q(z)$.

Prove that f is a holomorphic function on $\mathbb{P}^1(\mathbb{C})$.

Definition 4.1.3. Two Riemann Surfaces X, Y are called **isomorphic** (or **bi-holomorphic**) if there are holomorphic maps $f : X \to Y$ and $g : Y \to X$ such that $g \circ f = I_X$ and $f \circ g = I_Y$. In this case, we write $X \cong Y$ and call f and g **isomorphisms** (or **bi-holomorphisms**). An isomorphism $h : X \to X$ from a Riemann Surface to itself is called an **automorphism** of X.

Exercise 4.1.6. Let X, Y be Riemann Surfaces. Show that $X \cong Y$ (from Definition 4.1.3) if and only if there is a holomorphic map $f : X \to Y$ that is one-to-one and onto, and such that f^{-1} is holomorphic.

Exercise 4.1.7. Let $f : \mathbb{C} \to \mathbb{C}$ be a holomorphic function and $\Gamma_f \subset \mathbb{C}^2$ its graph (as defined in Example 3.1.2). Show that $\Gamma_f \cong \mathbb{C}$.

Exercise 4.1.8. In Exercise 3.2.11 you proved that any two smooth conics in $\mathbb{P}^2(\mathbb{C})$ are projectively equivalent; show that this implies that they are isomorphic as Riemann Surfaces.

Prove that any smooth conic is isomorphic to $\mathbb{P}^1(\mathbb{C})$ by choosing a particular conic for which you can exhibit an explicit isomorphism with the projective line.

4.2 Local Structure of Maps

An important feature of holomorphic maps of Riemann Surfaces is that they have many different local expressions near a point $x \in X$, depending on the choice of charts and local coordinates around x and $f(x)$; given a local coordinate function φ_x, post-composing with any bi-holomorphism h of $\mathbb{C}$ results in a new local coordinate function. For example, choosing $h : z \mapsto e^{(\pi/4)i}z$

rotates the coordinates by 45 degrees, or choosing $h : z \mapsto z + (2 + i)$ results in translated coordinates.

In fact, the map h doesn't need to be bi-holomorphic on all of $\mathbb{C}$; as long as it is bi-holomorphic "near" the point $\varphi_x(x)$, one can use $h \circ \varphi_x$ to get new coordinates around x.

For any holomorphic map f there exist choices of local coordinates such that the local expression of f around a given $x \in X$ is very simple: specifically, it is $z \mapsto z^k$ for some uniquely determined $k \geq 1$. This fact has remarkably deep implications in the theory of functions of Riemann Surfaces. We now make this statement precise.

We say that a chart (U_x, φ_x) for a Riemann Surface X is **centered at** x if $\varphi_x(x) = 0$.

Theorem 4.2.1. *Let $f : X \to Y$ be a non-constant holomorphic map of Riemann Surfaces. For any $x \in X$ there are charts centered at x, $f(x)$, such that the local expression of f using these charts is $z \mapsto z^k$ for some integer $k \geq 1$.*

Proof Start with any charts φ, ψ centered at x, $f(x)$, yielding a local expression denoted by $F := \psi \circ f \circ \varphi^{-1}$. Consider the Taylor expansion of F at 0 and let k be the smallest positive integer such that the coefficient of z^k does not vanish. Since $F(0) = 0, k \geq 1$ and:

$$F(z) = z^k \left(\sum_{n=0}^{\infty} a_{k+n} z^n \right). \tag{4.1}$$

Denote by $G(z) = a_k + a_{k+1} z + \cdots$ the second factor in (4.1). The function $G(z)$ is holomorphic at 0 and $G(0) = a_k \neq 0$. Thus we may make a choice of branch so that the map $\sqrt[k]{G(z)}$ is well-defined and holomorphic around $0 = \varphi(x)$.

Defining $h(z) = z\sqrt[k]{G(z)}$, we have that $F = h^k$. The function h is holomorphic in a neighborhood U of $0 = \varphi(x)$, $h(0) = 0$ and $h'(0) \neq 0$. The Inverse Function Theorem implies that h is bi-holomorphic on a neighborhood $U' \subset U$ of $\varphi(x)$, and therefore the composition $\tilde{\varphi} = h \circ \varphi$ gives a local chart for X centered at x.

The local coordinate $\tilde{z}$ coming from $\tilde{\varphi}$ is related to z via

$$\tilde{z} = h(z).$$

The local expression for f around x is now obtained by changing coordinates from z to $\tilde{z}$ in F:

$$\tilde{F}(\tilde{z}) = F(z(\tilde{z})) = h(z)^k = \tilde{z}^k. \tag{4.2}$$

□

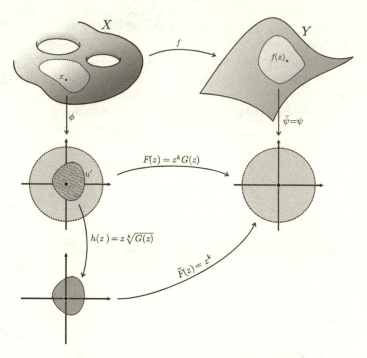

Figure 4.3 Altered charts to get local expression of z^k

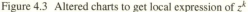
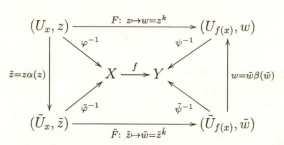

Figure 4.4 The ramification index is well-defined, i.e. independent of the choices of local coordinates that realize it

The next exercise shows that the integer k we associated to a map f and a point $x \in X$ is unique.

Exercise 4.2.1. Let $f : X \to Y$ be a non-constant holomorphic map of Riemann Surfaces, and $x \in X$. Suppose that f has two local expressions around x of the form $F(z) = z^k$ and $\tilde{F}(\tilde{z}) = \tilde{z}^{\tilde{k}}$. Then $k = \tilde{k}$.

Hint: Let $\varphi, \tilde{\varphi}, \psi, \tilde{\psi}$ be charts giving the appropriate coordinate functions, as illustrated in Figure 4.4. Observe that the change of coordinates near x must be of the form $\tilde{z} = z\alpha(z)$, where $\alpha(z)$ is holomorphic in a neighborhood of 0

and $\alpha(0) \neq 0$. Similarly (but note that we are going in the opposite direction), $w = \tilde{w}\beta(\tilde{w})$ for a holomorphic function β with $\beta(0) \neq 0$. Write down the Taylor expansions of α and β, and use the fact that F is obtained from $\tilde{F}$ via the above changes of coordinates to obtain the statement of the exercise.

Definition 4.2.2. Let $f : X \to Y$ be a non-constant holomorphic map of Riemann Surfaces. A picture is given in Figure 4.5.

- Given a point $x \in X$, the integer k_x, such that there exists a local expression centered at p of the form $F(z) = z^{k_x}$, is called the **ramification index** of f at x.
- The quantity $v_x = k_x - 1$ is called the **differential length** of f at x.
- If a point x has ramification index $k_x = 1$, then we say f is **unramified** at x.
- A point x such that $k_x \geq 2$ is called a **ramification point**. The **ramification locus** R is the subset of X consisting of all ramification points.
- If x is a ramification point, then $f(x) \in Y$ is called a **branch point**. The **branch locus** B is the subset of Y consisting of all branch points – i.e. the image via f of the ramification locus.

Warning! The branch locus is the image of the ramification locus, *but* the ramification locus is not necessarily the inverse image of the branch locus. (Why not?)

Remark 4.2.3. The function f is unramified at $x \in X$ (i.e. $k_x = 1$) if and only if for any local expression F of f around x (not necessarily centered at x) we have $F'(\varphi(x)) \neq 0$, i.e. f is locally invertible at x.

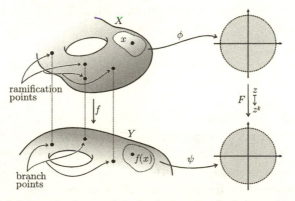

Figure 4.5 Schematic picture of the local structure of a holomorphic map of Riemann Surfaces

Exercise 4.2.2. Let $f : X \to Y$ be a non-constant holomorphic map of Riemann Surfaces, and $x_0 \in X$. Show that Remark 4.2.3 implies that there is a neighborhood U_0 of x_0 such that the ramification index of any $x \in U_0$ with $x \neq x_0$ is $k_x = 1$.

As an immediate consequence of Exercise 4.2.2 we have the following result.

Lemma 4.2.4. *The ramification locus R is a discrete subset of X, i.e. there exist open sets $U_i \subset X$ such that each U_i contains exactly one $x \in R$.*

A discrete subset of a compact topological space is finite; this is the contrapositive statement to "an infinite subset of a compact topological space has a limit point", which can be proven from the definition of compactness. As a consequence we have the following lemma.

Lemma 4.2.5. *If X is a compact Riemann Surface and $f : X \to Y$ is a non-constant holomorphic map of Riemann Surfaces, then the ramification locus is a finite set. Since the branch locus is the image of R via f, it follows that the branch locus is also a finite set.*

4.3 Maps of Compact Riemann Surfaces

From now on we restrict our attention to maps of compact Riemann Surfaces, where we also have some control over the global structure of the map. An important class of such functions to keep in mind is the meromorphic functions on a compact Riemann Surface X, i.e. holomorphic maps $f : X \to \mathbb{P}^1(\mathbb{C})$.

Theorem 4.3.1. *Let $f : X \to Y$ be a holomorphic map of Riemann Surfaces with X compact. If f is non-constant then it is onto.*

Proof   Assume that f is non-constant, and consider the image $f(X) \subseteq Y$: by Liouville's theorem it is an open set in Y (Conway, 1978, IV §7). On the other hand, since X is compact and f continuous, $f(X)$ is a compact subset of a Hausdorff topological space and therefore it is closed (Armstrong, 1983, Section 3.3, Theorem 3.6). Finally, since $f(X)$ is an open, closed and nonempty subset of a connected topological space, it follows that $f(X) = Y$.   □

The proof of Theorem 4.3.1 shows in particular that for f to be non-constant, Y must be compact as well. This immediately yields the following corollary.

Corollary 4.3.2. *If X is a compact Riemann Surface, the only holomorphic functions $f : X \to \mathbb{C}$ are constant functions.*

Let $f : X \to Y$ be a non-constant map of compact Riemann Surfaces, and consider a point x in the *fiber* (this is just a slightly fancier name for inverse image) $f^{-1}(y)$ of a point $y \in Y$. We know there exists a neighborhood U_x and appropriate choices of local coordinates centered at x such that the local expression F of f has form $z \mapsto z^{k_x}$. Since $F^{-1}(0) = 0$, there are no other preimages of y in U_x. This implies that the preimage set $f^{-1}(y)$ is discrete, and since X is compact, then (as in Lemma 4.2.5) it is a finite set. The next result shows that, outside of the branch locus, the size of the fibers of f, $|f^{-1}(y)|$, is constant.

Theorem 4.3.3. *Let $f : X \to Y$ be a non-constant holomorphic map of compact Riemann Surfaces. If $y_0, y_1 \in Y$ are not in the branch locus of f, then $|f^{-1}(y_0)| = |f^{-1}(y_1)|$.*

Proof Call B the branch locus of f. Since B is finite, $Y \setminus B$ is a connected topological space, and hence it cannot have a proper subset which is both open and closed. Let $y_0 \in Y \setminus B$ and set $d := |f^{-1}(y_0)|$.

The set $A = \{y \in Y \setminus B \,|\, |f^{-1}(y)| = d\}$ is nonempty, since it contains y_0. We claim A is open in $Y \setminus B$: for any $y \in A$, we show that y is an interior point of A. Denote by $x_1, \ldots, x_d$ the inverse images of y, and by $U_{x_1}, \ldots, U_{x_d}$ pairwise disjoint charts around each inverse image such that f admits local expression $F(z) = z$ on each chart. We defer to Exercise 4.3.1 to show that one may choose

$$V_y \subseteq f(U_{x_1}) \cap \ldots \cap f(U_{x_d}), \tag{4.3}$$

a connected open neighborhood of y, homeomorphic to a disc, such that the inverse image $f^{-1}(V_y)$ consists of d connected components $\tilde{U}_{x_i}$ (each containing one of the x_i and contained in U_{x_i}). Then f restricted to each $\tilde{U}_{x_i}$ is bijective onto V_y. This implies that every point of V_y has d inverse images, and hence that y is an interior point for A.

Exercise 4.3.1. Show that a V_y as in (4.3) exists, by showing that otherwise one may construct a sequence $\{\xi_n\}$ of points of X such that $\xi_n \to \bar{\xi}$, $f(\xi_n) \to y$ but $f(\bar{\xi}) \neq y$, violating the continuity of f.

Now consider $A^c = \{y \in Y \smallsetminus B \,||\, f^{-1}(y)| \neq d\}$: essentially with the same argument, one shows that A^c is also open in $Y \smallsetminus B$. This in turn implies that A is also closed.

Since A is a nonempty open and closed subset of a connected topological space, $A = Y \smallsetminus B$.   □

Theorem 4.3.3 makes the following definition well-defined.

Definition 4.3.4. For $f : X \to Y$ a non-constant holomorphic map of compact Riemann Surfaces, the **degree** of f is the cardinality of the fiber of any point y not in the branch locus of f. If f is constant we say that it has degree 0.

Exercise 4.3.2. Let $f : X \to Y$ be a holomorphic map of compact Riemann Surfaces of degree $d > 0$, $y \in Y$ and $f^{-1}(y) = \{x_1, \ldots, x_n\}$. Prove that

$$\sum_{i=1}^{n} k_{x_i} = d. \tag{4.4}$$

Exercise 4.3.2 generalizes Theorem 4.3.3 to say that the cardinality of any fiber of f is equal to the degree of f provided that points of X are weighted by their ramification index. Disregarding these multiplicities, this characterizes the branch locus of f as the set of points of Y that have strictly fewer than d preimages.

Remark 4.3.5. Exercise 4.3.2 implies that for f, a meromorphic function on X, the sum of the orders of zeroes of f equals the sum of the orders of poles of f.

4.4 The Riemann–Hurwitz Formula

The Riemann–Hurwitz formula gives a relation among all discrete invariants we have associated with a map of compact Riemann Surfaces.

Theorem 4.4.1 (Riemann–Hurwitz Formula). *Let $f : X \to Y$ be a non-constant, degree d, holomorphic map of compact Riemann Surfaces. Denote by g_X (respectively g_Y) the genus of X (respectively Y). Then*

$$2g_X - 2 = d(2g_Y - 2) + \sum_{x \in X} \nu_x, \tag{4.5}$$

where $\nu_x = k_x - 1$ is the differential length of f at x.

Since $v_x \neq 0$ if and only if x is a ramification point, the summation in (4.5) is a finite sum that may be equivalently indexed by points in the ramification locus.

Proof   Recall that the Euler Characteristic of a compact orientable surface of genus g is $2 - 2g$ (Section 2.4.2). Thus, the Riemann–Hurwitz Formula asserts that

$$\chi(X) = d\chi(Y) - \sum_{x \in X} v_x.$$

Given that we are comparing the Euler characteristics of X and Y, a natural strategy is to choose a suitable good graph on Y and "lift" it to a good graph on X which we use to compute $\chi(X)$.

Let Γ_Y be a good graph on Y with $B \subseteq V_{\Gamma_Y}$: the branch locus of f is contained in the vertex set of Γ_Y. Define Γ_X to be the "lift" of Γ_Y via the map f: the support of Γ_X is $f^{-1}(\Gamma_Y)$ and the vertices, edges and faces of Γ_X are the connected components of the inverse images of vertices, edges and faces of Γ_Y. See Figure 4.6.

After Chapter 5 we will be able to prove the following facts (see Remark 5.3.8): Γ_X is a good graph on X; the inverse image of each edge (respectively face) of Γ_Y consists of d distinct edges (respectively faces) of Γ_X.

For every vertex v of Γ_Y, the number of vertices of Γ_X mapping to v is precisely the number of inverse images of v. By Exercise 4.3.2

$$|f^{-1}(v)| = d - \sum_{\{x \mid f(x) = v\}} v_x,$$

which implies that $|V_{\Gamma_X}| = d|V_{\Gamma_Y}| - \sum_{x \in X} v_x$.

Computing $\chi(X)$ using Γ_X gives:

$$\begin{aligned}
\chi(X) &= |V_{\Gamma_X}| - |E_{\Gamma_X}| + |F_{\Gamma_X}| \\
&= d|V_{\Gamma_Y}| - \sum_{x \in X} v_x - d|E_{\Gamma_Y}| + d|F_{\Gamma_Y}| \\
&= d\chi(Y) - \sum_{x \in X} v_x.
\end{aligned}$$

$\square$

Example 4.4.2. Let $f : \mathbb{P}^1(\mathbb{C}) \to \mathbb{P}^1(\mathbb{C})$ be a degree d holomorphic map. Assume two points $x_1, x_2 \in \mathbb{P}^1(\mathbb{C})$ have full ramification, i.e. $k_{x_1} = k_{x_2} = d$; we use the Riemann–Hurwitz formula to show that there are no more ramification points for f. Setting $g_X = g_Y = 0$ in (4.5), we have

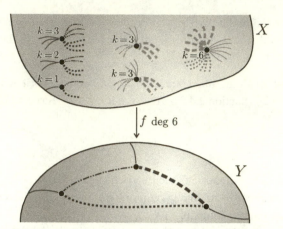

Figure 4.6 Lifting the graph on Y

$$-2 = d(-2) + (d-1) + (d-1) + \sum_{x \neq x_1, x_2} (k_x - 1)$$

$$0 = \sum_{x \neq x_1, x_2} \nu_x.$$

Since differential lengths are nonnegative integers, each $x \neq x_1, x_2$ must have $\nu_x = 0$.

Exercise 4.4.1. Consider $f : E \to \mathbb{P}^1(\mathbb{C})$, with $g_E = 1$. As in Example 4.4.2, assume there are two points $x_1, x_2 \in E$ where f is fully ramified. How many more ramification points can f have, and what ramification indices are possible?

Exercise 4.4.2. Suppose that $f : X \to Y$ is a non-constant holomorphic map of connected compact Riemann Surfaces. Prove that:

1. $\sum_{x \in X} \nu_x$ is even.
2. $g_X \geq g_Y$. A Riemann Surface X can never map (nontrivially) to a Riemann Surface Y of higher genus.
3. If $\sum_{x \in X} \nu_x = 0$ then $g_X = dg_Y - d + 1$.

Exercise 4.4.3. For readers who are familiar with differential forms, here is an alternative way to prove the Riemann Hurwitz formula:

1. Show that if ω is any meromorphic one form on a Riemann Surface X, then the sum of the orders or zeroes of ω minus the sum of the orders of poles is $2g_X - 2$.

2. Choose a meromorphic one form ω on Y such that the zeroes and poles of ω are disjoint from the branch locus of f, and pull it back via f. The differential form $f^*(\omega)$ is a meromorphic form on X. Prove that if x is a ramification point for x, then $f^*(\omega)$ has a zero of order v_x at x.
3. Compute the sum of orders of zeroes minus the sum of order of poles of $f^*(\omega)$ in two different ways to obtain the Riemann–Hurwitz formula.

4.5 Examples of Maps of Compact Riemann Surfaces

4.5.1 Maps from $\mathbb{P}^1(\mathbb{C})$ to $\mathbb{P}^1(\mathbb{C})$

We have seen in Exercise 4.1.5 that, given any two polynomials $p(z), q(z) \in \mathbb{C}[z]$ with no common roots, the rational function $f(z) = p(z)/q(z)$ defines a holomorphic map $f : \mathbb{P}^1(\mathbb{C}) \to \mathbb{P}^1(\mathbb{C})$, where we think of $\mathbb{P}^1(\mathbb{C})$ as $\mathbb{C} \cup \infty$. We now show that all holomorphic functions from the projective line to itself are of this form.

Theorem 4.5.1. *If $f : \mathbb{P}^1(\mathbb{C}) \to \mathbb{P}^1(\mathbb{C})$ is a holomorphic map of Riemann Surfaces, then f is a rational function: $f = p(z)/q(z)$, with $p(z), q(z) \in \mathbb{C}[z]$.*

Proof If f is a constant function then it is a rational function. Let f be a non-constant holomorphic function and denote by $z_1, \dots, z_n$ the inverse images of 0 and by $p_1, \dots, p_m$ the inverse images of ∞ via f. Define the rational function

$$\varphi(z) = \frac{\prod_{i=1}^{n}(z - z_i)^{k_{z_i}}}{\prod_{j=1}^{m}(z - p_j)^{k_{p_j}}},$$

where the k_{z_i} and k_{p_j} are the ramification indices of f at those points – which correspond to the orders of zeroes and poles of f in complex analysis. In the expression of φ we adopt the notation $(z - \infty) = 1$.

The function $f(z)/\varphi(z)$ does not take value 0 or ∞, and is therefore a non-surjective holomorphic function from $\mathbb{P}^1(\mathbb{C})$ to itself. By Theorem 4.3.1 it is constant. $\qquad\square$

Remark 4.5.2. If one prefers to describe points of $\mathbb{P}^1(\mathbb{C})$ via their homogeneous coordinates, then a holomorphic function $f : \mathbb{P}^1(\mathbb{C}) \to \mathbb{P}^1(\mathbb{C})$ is given as $f(X : Y) = (p(X, Y) : q(X, Y))$, where p and q are homogeneous polynomials in two variables of the same degree.

Exercise 4.5.1. Show that the descriptions of holomorphic functions in Theorem 4.5.1 and Remark 4.5.2 are equivalent.

We now consider which of these rational functions are automorphisms of $\mathbb{P}^1(\mathbb{C})$. Recall that an automorphism is a bijective function, so in particular it must have exactly one zero and one pole, implying that the degrees of $p(z)$ and $q(z)$ must be one.

Exercise 4.5.2. Let $f : \mathbb{P}^1(\mathbb{C}) \to \mathbb{P}^1(\mathbb{C})$ defined by $f(z) = (az+b)/(cz+d)$ for $a, b, c, d \in \mathbb{C}$. Show that f is an automorphism if and only if $ad - bc \neq 0$.

In complex analysis these maps f with $ad - bc \neq 0$ are called **Möbius transformations**.

Remark 4.5.3. Automorphisms of $\mathbb{P}^1(\mathbb{C})$ can also be described using homogeneous coordinates for $\mathbb{P}^1(\mathbb{C})$. This amounts to thinking of $\mathbb{P}^1(\mathbb{C}) = (\mathbb{C}^2 \smallsetminus \{\vec{0}\})/\mathbb{C}^*$ as the space of one-dimensional linear subspaces of $\mathbb{C}^2$. It is natural then to obtain an automorphism of $\mathbb{P}^1(\mathbb{C})$ from a linear automorphism of $\mathbb{C}^2$, i.e. an element $M \in \mathrm{GL}(2, \mathbb{C})$ where

$$\mathrm{GL}(2, \mathbb{C}) = \left\{ \left(\begin{array}{cc} a & b \\ c & d \end{array} \right) \middle| a, b, c, d \in \mathbb{C} \text{ with } ad - bc \neq 0 \right\}.$$

Identifying a point $(X : Y) \in \mathbb{P}^1(\mathbb{C})$ with a column vector, the action is given by matrix multiplication. Note that two matrices M_1 and M_2 yield the same map on projective space if their entries differ by scaling by a nonzero constant. The group of automorphisms of $\mathbb{P}^1(\mathbb{C})$ is therefore naturally identified with $\mathbb{P}\mathrm{GL}(2, \mathbb{C})$, the **projective general linear group**.

Exercise 4.5.3. Prove that there is an isomorphism of groups between the group of Möbius transformations, with operation composition of functions, and $\mathbb{P}\mathrm{GL}(2, \mathbb{C})$, with operation matrix multiplication.

4.5.2 Maps of Elliptic Curves

Consider an elliptic curve, i.e. a projective algebraic plane curve defined as $E = V(P)$, where P is a polynomial of the form:

$$P(X, Y, Z) = Y^2 Z - (X - a_1 Z)(X - a_2 Z)(X - a_3 Z).$$

We have seen in Example 3.2.8 that E is a smooth curve (i.e. a Riemann Surface) if and only if the a_is are distinct complex numbers. We now use the Riemann Hurwitz formula to determine the genus of E.

Lemma 4.5.4. *An elliptic curve is a Riemann Surface of genus 1.*

Proof Consider the affine chart $U_Z = \{Z \neq 0\} \subseteq \mathbb{P}^2(\mathbb{C})$, with coordinates $(x, y) = (\frac{X}{Z}, \frac{Y}{Z})$. The restriction of E to this chart is the affine curve E_Z determined by the equation $y^2 = (x - a_1)(x - a_2)(x - a_3)$.

The vertical projection map $\pi : (x, y) \mapsto x$ restricts to a holomorphic map $\pi : E_Z \to \mathbb{C}$. One can verify that for every point x of $\mathbb{C}$ except for the a_i's, π^{-1} consists of two points, i.e. the degree of π is equal to 2.

It is a consequence of the Riemann Existence Theorem (Theorem 6.2.2) that π extends to a holomorphic map $\tilde{\pi} : E \to \mathbb{P}^1(\mathbb{C})$ by $(0 : 1 : 0) \mapsto \infty$. The result is a map of degree 2 of connected compact Riemann Surfaces.

The branch locus for $\tilde{\pi}$ is $B = \{a_1, a_2, a_3, a_4 = \infty\}$; denote r_1, r_2, r_3, r_4 as the corresponding ramification points. Since for a map of degree 2 the only nontrivial ramification has differential length equal to 1, the Riemann–Hurwitz formula reads:

$$2g_E - 2 = 2(-2) + \sum_{r_1, r_2, r_3, r_4} 1, \tag{4.6}$$

which gives $g_E = 1$. □

In section 3.2.2 we encountered Riemann Surfaces of genus 1: complex tori. It is a beautiful and classical story (see, for example, Silverman and Tate (1992)) that for any complex torus T, there is an equation which realizes T as an elliptic curve $E \cong T$, and vice versa that any elliptic curve is isomorphic to a complex torus. The rich and tight interplay between the algebraic and complex analytic points of view is one of the interesting and fruitful features of the theory of Riemann Surfaces. In the next two exercises we describe maps among elliptic curves by taking the complex tori point of view, where the description becomes simple.

Exercise 4.5.4. A non-constant holomorphic map between complex tori $\tilde{f} : \mathbb{C}/\Lambda \to \mathbb{C}/\Lambda'$ is called an **isogeny**. An isogeny is defined by a holomorphic map $f : \mathbb{C} \to \mathbb{C}$ which is well-defined when we mod out by the lattices, i.e. such that for any $z \in \mathbb{C}$ and any $l \in \Lambda$ we have $f(z + l) = f(z) + l'$ for some $l' \in \Lambda'$.

1. Use the Riemann–Hurwitz formula to show that any isogeny f is unramified.
2. Show that any isogeny is induced by a map $f(z) = az$ where $a \in \mathbb{C}$ and $f(\Lambda) \subset \Lambda'$.
3. Consider the isogeny $\tilde{f} : \mathbb{C}/\Lambda \to \mathbb{C}/\Lambda'$ induced by $f(z) = z + 1$, where $\Lambda = \{n + m(1+i) | n, m \in \mathbb{Z}\}$ and $\Lambda' = \{n(1/2) + m(1/2 + i/2) | n, m \in \mathbb{Z}\}$. Find the degree of $\tilde{f}$.

Exercise 4.5.5. Consider a lattice Λ generated by the vectors $\rho_1 e^{i\theta_1}$, $\rho_2 e^{i\theta_2}$, with $0 \leq \theta_1 < \theta_2 < 2\pi$. Denote by $\tau = \frac{\rho_2}{\rho_1} e^{i(\theta_2 - \theta_1)}$ and by Λ' the lattice generated by the vectors 1 and τ. Prove that the complex tori $\mathbb{C}/\Lambda$ and $\mathbb{C}/\Lambda'$ are isomorphic.

5

Loops and Lifts

Let us begin this chapter with a somewhat facetious metaphor: given a degree 3 map of Riemann Surfaces $f : X \to Y$ with branch locus B, imagine $Y \setminus B$ is the world you live in and the map f gives the vertical projection from the heavens $X \setminus f^{-1}(B)$. Chapter 4 tells us that an astronomer[1], pointing the telescope straight up into the sky, observes that the portion of heavens seen is always three copies of what surrounds her. A first guess is that the heavens are then exactly three copies of the Earth, but she suspects that there may be other possibilities, and devises the following experiment to test such theory. Standing at a particular point on Earth she finds a way to "mark" a, b, c the three points in the sky lying precisely above her head. Keeping her focus on the portion of sky identified by a, she starts walking around while looking into the telescope. If the heavens are indeed three copies of Earth, every time she comes back to the original point, her gaze should return to a. So if she finds one particular walk such that when she returns she is looking at b or c, this experiment will show that the global geometry of the heavens is in fact different from three copies of Earth.

This chapter is devoted to turning this silly metaphor into actual mathematics. Our goal is to introduce the notion of *coverings*, i.e. pairs of spaces whose local geometry is identical; the global geometries are then controlled by the groups of loops of the two spaces. We begin by introducing the notion of homotopy of functions, corresponding to the idea of "wiggling" one function into another. We define the fundamental group of a pointed topological space as the group of homotopy equivalence classes of loops originating at the base point of the topological space. We show that, given a covering, the fundamental group of the source space is naturally identified with a subgroup of

---

[1] We assume our astronomer is female in honor of great astrophysicist Margerita Hack (1922–2013), even though (as far as we know) she never lived inside a Riemann Surface.

the fundamental group of the target space, and that in fact there is a perfect "dictionary" between coverings of a given (pointed) space and subgroups of its fundamental group, known as the *Galois Correspondence of Covering Theory*.

As with other background chapters in this book, our treatment of this subject is somewhat skeletal and very much tuned to what we use in the rest of the book. Any basic topology textbook (e.g. Armstrong (1983) and Munkres (1975)), or the very friendly graduate textbook (Hatcher, 2002), may be consulted by the reader who is seeking a more comprehensive treatment.

5.1 Homotopy

The notion of homotopy formalizes the idea of continuously altering (or "wiggling") maps of topological spaces. A picture is given in Figure 5.1.

The intuition is that a homotopy between two maps f and g is a one-hour-long movie, which starts showing the map f, and ends with the map g.

Definition 5.1.1. Let $f, g : X \to Y$ be maps of topological spaces.

- A **homotopy** between f and g is a continuous function

$$H : X \times [0, 1] \to Y$$

  such that $H(x, 0) = f(x)$ and $H(x, 1) = g(x)$ for all $x \in X$.
- If a homotopy H exists, we say that f and g are **homotopic** and write $f \sim g$.
- Let $A \subseteq X$ be such that $f_{|A} = g_{|A}$. A homotopy H between f and g is said to be **relative to** A if, for every $a \in A$ and for every $t \in [0, 1]$, we have that $H(a, t) = f(a) = g(a)$.

Example 5.1.2. For any topological space X, any two continuous maps $f, g : X \to \mathbb{R}^n$ are homotopic. This can be seen by considering the *straight-line homotopy* $H : X \times [0, 1] \to \mathbb{R}^n$ given by

$$H(x, t) = (1 - t)f(x) + tg(x).$$

Example 5.1.2 shows that the topology of Euclidean spaces allows any two functions to be deformed into one another in the simplest possible way: by just moving from any point of the form $f(x)$ to the corresponding point $g(x)$ along the straight line segment joining the two points. For a different target, it therefore seems reasonable that the failure to execute this strategy might be an indication of some interesting geometric feature of the target space. The next

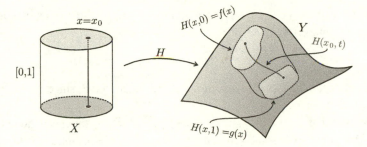

Figure 5.1 Schematic picture of a homotopy of maps

example concerning homotopies of paths illustrates this point (a *path* in Y is a continuous map $\gamma : [0, 1] \to Y$).

Example 5.1.3. Consider the functions $f, g : [0, 1] \to \mathbb{R}^2 \smallsetminus (0, 0)$ defined by:

$$f(s) = (\cos(2\pi s), \sin(2\pi s)), \quad g(s) = (1, 0).$$

Let $A = \{0, 1\}$ consist of the endpoints of the segment. We prove that f and g are not homotopic relative to A. We first observe that the naive attempt of using the straight line homotopy fails because we have removed the point $(0, 0)$ from the target space, and this prevents the point $(-1, 0)$ from moving in a straight line until it reaches $(1, 0)$. Intuitively, it seems reasonable that we can't "pull in" the unit circle to a constant loop without crossing the origin.

We now provide two formal proofs of this fact: we first go at it with as little technology as possible, and notice that the proof is considerably more involved than the statement seems to call for. Next, we call on complex analysis for help and uncover a much slicker proof. With this we illustrate the philosophy that learning mathematics that is more advanced than the problem at hand can be a valuable asset.

Proof 1: bare hands. Draw a picture as you read this proof to follow the details.

Assume there exists H a homotopy relative to A between f and g

$$H : [0, 1] \times [0, 1] \to \mathbb{R}^2 \smallsetminus (0, 0).$$

Denote:

$$U^+ := H^{-1}(\{y > 0\}) \quad U^- := H^{-1}(\{y < 0\}) \quad K^0 := H^{-1}(\{y = 0\}).$$

We note that three sides of the square other than the $\{t = 0\}$ side are contained in K^0: H restricted to these sides is the constant map $(1, 0)$. Since $(0, 1/2) \times \{0\} \subseteq U^+$ and $(1/2, 1) \times \{0\} \subseteq U^-$, there must exist a path γ in $[0, 1] \times [0, 1]$

such that $\gamma(0) = (1/2, 0)$, $\gamma(1)$ belongs to one of the three sides mentioned above, and $\text{Image}(\gamma) \subseteq K^0$. Define

$$\phi = H \circ \gamma : [0, 1] \to \mathbb{R} \smallsetminus 0 \subset \mathbb{R}^2 \smallsetminus (0, 0).$$

We may think of ϕ as a function taking values in $\mathbb{R}$ since its image is contained in the x-axis.

We have that $\phi(0) = -1$ and $\phi(1) = 1$; therefore, by the mean value theorem, there must exist some value of t for which $\phi(t) = 0$. But this contradicts the assumption that $(0, 0)$ does not belong to the image of H.

Proof 2: with complex analysis. Identify $\mathbb{R}^2$ with $\mathbb{C}$ and notice that f and g define paths in the complex plane. Remember, in Theorem 1.2.2 we saw that path integrals of holomorphic functions are invariant under (now we can say the appropriate word!) endpoint-preserving homotopies of paths. Consider the function $1/z$, which is holomorphic on $\mathbb{C} \smallsetminus 0$. In Example 1.2.1 we computed

$$\int_f \frac{1}{z} dz = 2\pi i.$$

On the other hand, the path integral of any function along the constant path g must be 0. It follows that f cannot be homotopic to g.

Exercise 5.1.1. Prove that the constant map $\pi : S^1 \to S^1$ defined by $\pi(x) = [0]$ is *not* homotopic to Id_{S^1}. You may think of S^1 as the topological space $[0, 1]/0 \sim 1$ and use Example 5.1.3 to solve this exercise.

Exercise 5.1.2. Prove that any two continuous maps $f, g : \{pt\} \to X$ are homotopic if and only if X is path-connected. Here $\{pt\}$ is a topological space with only one point.

Exercise 5.1.3. Show that homotopy defines an equivalence relation on the set of maps from a topological space X to a topological space Y.

We use our newly defined notion of equivalence of continuous functions to define a notion of equivalence of topological spaces that relaxes the notion of homeomorphism.

Definition 5.1.4. Two topological spaces X, Y are called **homotopy equivalent** (or simply **homotopic**) and denoted $X \sim Y$, if there exist maps $f : X \to Y$ and $g : Y \to X$ such that

$$g \circ f \sim Id_X, \quad f \circ g \sim Id_Y.$$

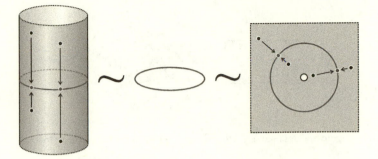

Figure 5.2 The cylinder, circle and punctured plane are homotopy equivalent

Example 5.1.5. A topological space which is homotopy equivalent to a point is called **contractible**. We show that any Euclidean space is contractible: for any value of n, $\mathbb{R}^n \sim \{pt\}$.

Consider the constant map $\pi : \mathbb{R}^n \to \{pt\}$ and the inclusion $\iota : \{pt\} \to \mathbb{R}^n$ given by $\iota(pt) = \vec{0}$. We have $\pi \circ \iota = Id_{pt}$, and $\iota \circ \pi : \mathbb{R}^n \to \mathbb{R}^n$ is the constant function $\vec{0}$. The latter composition is homotopic to $Id_{\mathbb{R}^n}$, for example via a straight line homotopy as in Example 5.1.2.

Exercise 5.1.4. Show, using Figure 5.2 as a guide, that the cylinder, the punctured plane and the circle are homotopy equivalent. Choose wisely how to give coordinates to points of each of these spaces.

5.2 The Fundamental Group

The *fundamental group* is a topological invariant which, to a topological space X, associates a group of equivalence classes of loops on X up to wiggling. In this section we collect the basic definitions and highlight the features of this theory that are important to us.

Definition 5.2.1. Let X be a topological space and $x_0 \in X$. A **loop in X with base point x_0** is a continuous map $\gamma : [0, 1] \to X$ such that $\gamma(0) = \gamma(1) = x_0$.

Two loops γ, δ with base point x_0 are said to be **homotopic with respect to the base point** (see Figure 5.3) if there exists a homotopy $H : [0, 1] \times [0, 1] \to X$ between γ and δ such that for every $t \in [0, 1]$ we have $H(0, t) = H(1, t) = x_0$.

Being homotopic with respect to the base point is an equivalence relation on the set of loops based at x_0. If γ and δ are equivalent loops, we write $\gamma \sim \delta$.

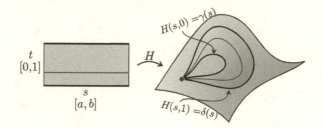

Figure 5.3 A homotopy with respect to a base point between two loops

It is important to stress that loops are functions to X, not subsets of X. If one thinks of t as a time variable, a loop is a motion of a particle in X, not just its trajectory. *If you like to jog, you should think of a loop as your one-hour-long morning run, not just as your running route.* Given two loops γ_1, γ_2, intuitively one can get a new loop by first "running around" γ_1 and then "running around" γ_2. Since loops must have domain $[0, 1]$ we are now forced to run twice as fast. We formalize this idea into a binary operation on the set of based loops.

Definition 5.2.2. Given two loops γ_1, γ_2 in X with base point x_0, we define the loop $\gamma_1 * \gamma_2$ in X with base point x_0 as follows:

$$\gamma_1 * \gamma_2(s) = \begin{cases} \gamma_1(2s) & \text{if } s \in [0, 1/2] \\ \gamma_2(2s - 1) & \text{if } s \in [1/2, 1]. \end{cases}$$

Lemma 5.2.3. *The operation of concatenation of loops $*$ from Definition 5.2.2 is compatible with homotopy equivalence of based loops: if $\gamma_1 \sim \delta_1$ and $\gamma_2 \sim \delta_2$, then*

$$\gamma_1 * \gamma_2 \sim \delta_1 * \delta_2.$$

Proof   Let H_1 be a homotopy between γ_1 and δ_1, and similarly for H_2 between γ_2 and δ_2, both with respect to the base point x_0. Then we create a homotopy H between $\gamma_1 * \gamma_2$ and $\delta_1 * \delta_2$ with respect to x_0 by "placing H_1 and H_2 side by side". Specifically, we define $H : [0, 1] \times [0, 1] \rightarrow X$ as follows (see Figure 5.4):

$$H(s, t) = \begin{cases} H_1(2s, t) & \text{if } s \in [0, 1/2] \\ H_2(2s - 1, t) & \text{if } s \in [1/2, 1]. \end{cases}$$

□

Theorem 5.2.4. *Let X be a topological space and $x_0 \in X$. Then the set of equivalence classes of loops with basepoint x_0 is a group under the binary operation $*$.*

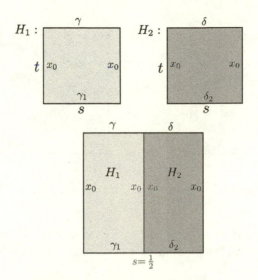

Figure 5.4 Concatenating two homotopies

Proof Lemma 5.2.3 implies that $*$ induces an operation on the quotient set: given two equivalence classes of based loops $[\gamma_1]$, $[\gamma_2]$,

$$[\gamma_1] * [\gamma_2] := [\gamma_1 * \gamma_2].$$

Having a well-defined binary operation, we must show that the group axioms hold.

Associativity The equivalence:

$$(\gamma_1 * \gamma_2) * \gamma_3 \sim \gamma_1 * (\gamma_2 * \gamma_3)$$

is given by the homotopy:

$$H(s,t) = \begin{cases} \gamma_1(\frac{4s}{t+1}) & \text{if } t \geq 4s - 1 \\ \gamma_2(4s - (t+1)) & \text{if } 4s - 2 \leq t \leq 4s - 1 \\ \gamma_3(\frac{4s-t-2}{2-t}) & \text{if } t \leq 4s - 2. \end{cases} \qquad (5.1)$$

Identity The identity element e with respect to $*$ is the (class of the) constant loop $\epsilon_{x_0}(s) = x_0$ for all $s \in [0, 1]$. The identity:

$$[\epsilon_{x_0}] * [\gamma] = [\gamma]$$

is shown by the homotopy

$$H(s,t) = \begin{cases} \gamma(\frac{2s}{t+1}) & \text{if } t \geq 2s - 1 \\ x_0 & \text{if } t \leq 2s - 1. \end{cases} \qquad (5.2)$$

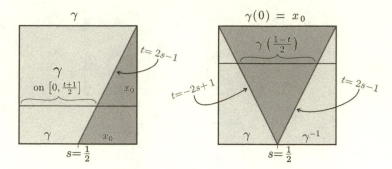

Figure 5.5 Homotopy squares for two of the group axioms

Inverse Given an equivalence class $[\gamma]$, its inverse is $[\gamma]^{-1} = [\gamma^{-1}]$, where $\gamma^{-1}(s) = \gamma(1-s)$. Note that γ^{-1} "walks backwards" along γ. To show that

$$[\gamma^{-1}] * [\gamma] = [\epsilon_{x_0}],$$

consider the homotopy

$$H(s,t) = \begin{cases} \gamma(s) & \text{if } s \leq \frac{1-t}{2} \\ \gamma(\frac{1-t}{2}) & \text{if } \frac{1-t}{2} \leq s \leq \frac{t+1}{2} \\ \gamma^{-1}(s) & \text{if } \frac{t+1}{2} \leq s. \end{cases} \tag{5.3}$$

The homotopies in (5.2) and (5.3) are represented in Figure 5.5. We leave it to Exercise 5.2.1 to finish the check that the identity and inverse axioms hold, and hence to complete the proof of Theorem 5.2.4.   □

Exercise 5.2.1. Draw the homotopy representing (5.1) and write out explicit homotopies to show the following:

1. $[\gamma] * [\epsilon_{x_0}] = [\gamma]$;
2. $[\gamma] * [\gamma^{-1}] = [\epsilon_{x_0}]$.

After Theorem 5.2.4, we can finally make the following definition.

Definition 5.2.5. Let X be a topological space and $x_0 \in X$. The **fundamental group** of X with base point x_0, denoted $\pi_1(X, x_0)$, is the group of equivalence classes of loops based at x_0, with operation induced by concatenation of loops$*$.

Given $f : X \to Y$ a continuous map, we now define a function $\pi_1(f)$ between the fundamental groups:

$$\pi_1(f): \quad \pi_1(X, x_0) \quad \to \quad \pi_1(Y, f(x_0))$$
$$[\gamma] \quad \mapsto \quad [f \circ \gamma]. \tag{5.4}$$

Exercise 5.2.2. Let $f : X \to Y$ and $g : Y \to Z$ be continuous maps sending $x_0 \mapsto y_0 \mapsto z_0$. Prove that:

(A1) $\pi_1(f) : \pi_1(X, x_0) \to \pi_1(Y, y_0)$ is a group homomorphism;
(A2) $\pi_1(Id_X) = Id_{\pi_1(X, x_0)}$;
(A3) $\pi_1(g \circ f) = \pi_1(g) \circ \pi_1(f)$.

Remark 5.2.6 (Aside on categories and functors). Category theory argues that any field of mathematics consists of two ingredients: a collection of *objects* that one is interested in studying, and a class of *good functions* that allow one to relate different objects. We call a choice of these ingredients a *category*. Thus group theory, for example, is the study of the category of groups $\mathcal{G}$, whose objects are groups and whose good functions are group homomorphisms. Topology is the study of the category $\mathcal{T}$ of topological spaces and continuous functions. Then we may consider the variant $\mathcal{PT}$, the category of *pointed* topological spaces: objects are pairs (X, x_0), where X is a topological space and $x_0 \in X$; a good function $f : (X, x_0) \to (Y, y_0)$ is a continuous function such that $f(x_0) = y_0$.

Taking this point of view, a natural question is what the notion of a good function among categories should be. Rather than getting sidetracked with abstract theory for too much longer (the reader interested in some details may refer to Hilton and Stammbach (1997)), we say that a good function of categories is called a *functor* and point out that the fundamental group offers us the first example of this concept. The fundamental group functor $\pi_1 : \mathcal{PT} \to \mathcal{G}$ assigns a group to every pointed topological and a group homomorphism between the corresponding groups to any function of pointed topological spaces. Further, such assignment preserves the notion of identity function ((A2) in Exercise 5.2.2) and respects composition of functions ((A3) in Exercise 5.2.2).

The formal properties of a functor allow us to prove in a standard way that the fundamental group is a topological invariant.

Proposition 5.2.7. *Let* (X, x_0), (Y, y_0) *be homeomorphic pointed topological spaces. Then* $\pi_1(X, x_0) \cong \pi_1(Y, y_0)$.

Proof Note that (X, x_0) is homeomorphic to (Y, y_0) if and only if there are continuous maps $f : X \to Y$ and $g : Y \to X$ such that

$$g \circ f = Id_X, \qquad f \circ g = Id_Y, \qquad f(x_0) = y_0.$$

Applying π_1 to the first identity and then using (A2) and (A3) gives

$$\pi_1(g) \circ \pi_1(f) = \pi_1(g \circ f) = \pi_1(Id_X) = Id_{\pi_1(X, x_0)}. \qquad (5.5)$$

The second equality, $f \circ g = Id_Y$, similarly yields $\pi_1(f) \circ \pi_1(g) = Id_{\pi_1(Y, y_0)}$. Thus $\pi_1(f)$ and $\pi_1(g)$ are inverses of each other and $\pi_1(X, x_0) \cong \pi_1(Y, y_0)$. $\qquad\qquad\qquad\qquad\qquad\qquad\qquad\qquad\qquad\qquad\qquad\qquad\qquad$ $\square$

Remark 5.2.8. The information of a base point is often unnecessary if one is only interested in the isomorphism class of fundamental groups. Precisely, if X is path-connected and $x_0, x_1 \in X$, then $\pi_1(X, x_0) \cong \pi_1(X, x_1)$ (and sometimes we just write $\pi_1(X)$). The isomorphism is non-canonical, as it relies on the choice of a path σ from x_1 to x_0. For any such σ, the function $\Phi_\sigma : \pi_1(X, x_0) \to \pi_1(X, x_1)$ defined by $[\gamma] \mapsto [\sigma^{-1} * \gamma * \sigma]^2$ is a group isomorphism.

In fact, fundamental groups are very loose invariants. They are not only constant in homeomorphism classes but also in homotopy classes.

Exercise 5.2.3. Suppose that X, Y are path-connected and $f, g : X \to Y$ are continuous maps. Show that if f is homotopic to g, then $\pi_1(f) = \pi_1(g)$. Use this fact to show that if X is homotopy equivalent to Y, then $\pi_1(X) \cong \pi_1(Y)$.

5.2.1 Examples

Computing fundamental groups is typically a nontrivial task. There are a few standard tools that one studies in a first course in topology which, used in combination, allow us to compute the fundamental groups of most of the geometric objects that we encounter in this book. We refer the reader interested in a more in-depth treatment of the topic to Armstrong (1983), Munkres (1975) and Hatcher (2002). In this section we present a collection of facts and examples, without proof but with some motivation, with the goal of developing a working understanding of the theory sufficient for our goals.

Contractible Spaces. Since $\pi_1(\{pt\}) = \{e\}$, it follows that X contractible implies $\pi_1(X) = \{e\}$.

---

[2] Here $*$ denotes the associative operation of concatenation of paths, which is defined analogously to the operation of concatenation of loops.

Sphere. The sphere S^2 is not contractible, but any loop on the sphere can be "pulled in", i.e. is homotopic[3] to ϵ_{x_0}. Thus $\pi_1(S^2) = \{e\}$. Any path-connected space X such that $\pi_1(X) = \{e\}$ is called **simply connected**. For any $n \geq 2$, the n-dimensional sphere S^n is simply connected.

Circle. Any loop on the circle S^1 is homotopic to a loop that travels at constant speed around the circle clockwise (or counterclockwise) an integer number of times. Furthermore, each loop can be seen as multiple iterations of a loop which travels around only once. Thus

$$\pi_1(S^1) = \mathbb{Z},$$

or equivalently the free group on 1 generator, $\mathbb{F}_1$.

Flower graph. Take g circles, label a point on each one, and glue them together at the chosen points. We call the resulting space a *flower graph* (see the right-hand side of Figure 5.6). The loops on such a graph are generated by the simple loops which go around each "petal" once, and there are no relations between them. Thus, if the graph Γ has g petals, then $\pi_1(\Gamma) = \mathbb{F}_g$, the free group on g generators.

Graphs. Let Γ be a connected graph. By contracting a *spanning tree*[4] on Γ we see that Γ is homotopy equivalent to a flower graph with a certain number of petals, g (see Figure 5.6). The number g is the genus of the graph and is given by the formula $1 - g = \chi(\Gamma) = V - E$, where V is the number of vertices of Γ and E the number of edges. Thus

$$\pi_1(\Gamma) = \mathbb{F}_g = \mathbb{F}_{E-V+1}. \tag{5.6}$$

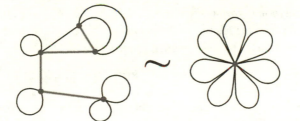

Figure 5.6 Contracting this spanning tree yields a flower graph with seven petals

---

[3] This is not as trivial as it may seem: it is possible for a continuous loop to go through every point of the sphere, and one has to show that such a loop is homotopic to a non-surjective loop; the latter can then easily be retracted. One honest possibility for formally proving that S^2 is simply connected is to apply the Seifert–Van Kampen theorem.

[4] A spanning tree for a graph Γ is a subgraph of Γ which contains all vertices of Γ and does not contain any cycles.

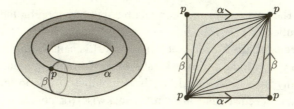

Figure 5.7 A homotopy of loops on the torus. Try and picture how the loops in the homotopy on the right-hand side look on the surface of the torus

Punctured Sphere. A sphere with one point removed is contractible, so $\pi_1(S^2 \smallsetminus \{p\}) = \{e\}$. A sphere with two points removed is homotopy equivalent to a circle, so $\pi_1(S^2 \smallsetminus \{p_1, p_2\}) = \mathbb{F}_1$. A sphere with n points removed is homotopy equivalent to a flower graph with $n - 1$ petals, so

$$\pi_1(S^2 \smallsetminus \{p_1, \ldots, p_n\}) = \mathbb{F}_{n-1}.$$

Torus. Let T be the torus defined as the identification space $T = [0, 1] \times [0, 1]/ \sim$ where we identify $(0, y) \sim (1, y)$ and $(x, 0) \sim (x, 1)$. There are two fundamental loops α, β, parameterizing the coordinate segments on T, and they satisfy a relation: Figure 5.7 illustrates a homotopy between the loops $\beta * \alpha$ and $\alpha * \beta$. Thus

$$\pi_1(T) \cong \langle a, b | aba^{-1}b^{-1} \rangle \cong \mathbb{Z} \oplus \mathbb{Z}.$$

Orientable Surfaces. A general surface T_g of genus g may be obtained as an identification space by identifying appropriately pairs of sides of a $4g$-gon, as in Section 2.4.1. Each of the sides of the polygon becomes a loop after the identification, and a path parameterizing the boundary of the polygon corresponds to a path that can be contracted to a constant path through the interior of the polygon. This gives rise to the following presentation for the fundamental group:

$$\pi_1(T_g) \cong \langle a_1, b_1, \ldots a_g, b_g | a_1 b_1 a_1^{-1} b_1^{-1} \ldots a_g b_g a_g^{-1} b_g^{-1} \rangle. \quad (5.7)$$

Exercise 5.2.4. Prove that the space obtained by removing n points from a genus g orientable surface is homotopy equivalent to a flower graph with $2g + n - 1$ petals. Show also that the fundamental group of such a punctured surface can be presented in a way which is symmetric with respect to all the punctures:

$$\pi_1(T_g \smallsetminus \{p_1, \ldots, p_n\})$$
$$\cong \langle a_1, b_1, \ldots a_g, b_g, r_1, \ldots, r_n | a_1 b_1 a_1^{-1} b_1^{-1} \ldots a_g b_g a_g^{-1} b_g^{-1} r_1 \ldots r_n \rangle.$$
$$(5.8)$$

5.3 Covering Spaces

Informally, a covering $X \to Y$ is a surjective map which is a local homeomorphism at every point. This means that for any $y \in Y$, the local geometry around y is the same as the local geometry around each inverse image of y (see Figure 5.8). An elementary example of a covering, also called a **trivial cover**, is when X consists of a number of disjoint copies of Y and the map restricted to each copy is just the identity function. However, there are more interesting examples of coverings, and the global geometry of coverings of Y is tightly connected to the fundamental group of Y.

Definition 5.3.1. A **covering** is a continuous, surjective map $p : X \to Y$ such that for every $y \in Y$ and each $x_i \in p^{-1}(y)$ there exists a neighborhood U_y of y whose inverse image $p^{-1}(U_y)$ consists of disjoint neighborhoods V_{x_i} and each restriction of p to V_{x_i} is a homeomorphism $p : V_{x_i} \to U_y$.

For any covering $p : X \to Y$ and $y \in Y$ we have that $p^{-1}(y)$ is a discrete set. If X is connected and locally path-connected, the proof of Theorem 4.3.3 naturally generalizes to show that $|p^{-1}(y)|$ is the same for all $y \in Y$ – in this case we call $|p^{-1}(y)|$ the **degree** of p. In Chapter 4 we saw that maps of compact Riemann Surfaces provide many examples of coverings.

Example 5.3.2. Let $f : X \to Y$ be a holomorphic map of compact Riemann Surfaces with ramification locus $R \subset X$ and branch locus $B \subset Y$. Then the restriction $f : (X \smallsetminus R) \to (Y \smallsetminus B)$ is a covering of finite degree.

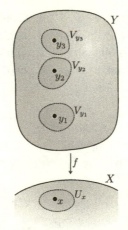

Figure 5.8 Schematic picture of a covering

Example 5.3.3. Identify the circle S^1 with the complex unit circle $S^1 = \{z \in \mathbb{C} | |z| = 1\}$.

- The map $p : S^1 \to S^1$ defined by $p(z) = z^4$ is a covering of degree 4. One similarly has a covering for each map $p(z) = z^d$ where $d \in \mathbb{N}$.
- Define the map $q : \mathbb{R} \to S^1$ by $q(t) = e^{2\pi i t}$. Then q is a covering with each set $q^{-1}(z)$ in bijection with $\mathbb{Z}$.

Coverings and fundamental groups are intimately connected. We will show that the natural map $\pi_1(p) : \pi_1(X, x_0) \to \pi_1(Y, y_0 = p(x_0))$ is injective, which allows us to think of $\pi_1(X, x_0)$ as a subgroup of the fundamental group of Y. Conversely, given any loop γ in Y based at y_0, there always exists a unique path $\tilde{\gamma}$ starting at x_0 which projects to γ via p. It is important to notice that $\tilde{\gamma}$, which is called a lift of γ, need not be a loop, as it may end at a different preimage of y_0. In fact, one may think of (the image of) $\pi_1(X, x_0)$ as the subgroup of $\pi_1(Y, y_0)$ consisting of loops which lift to loops.

Definition 5.3.4. Given a covering $p : X \to Y$ and a continuous function $\alpha : B \to Y$, a **lift** of α is a continuous function $\tilde{\alpha} : B \to X$ such that $p \circ \tilde{\alpha} = \alpha$.

When liftings exist, they are "almost" unique. We make this statement precise in the following exercise.

Exercise 5.3.1. Given a covering $p : X \to Y$, and two lifts $\tilde{\alpha}_1, \tilde{\alpha}_2$ of a continuous function $\alpha : B \to Y$, with B connected. Then either $\tilde{\alpha}_1 = \tilde{\alpha}_2$ or the images of the two lifts are disjoint. Show that both situations may occur. **Hint:** *consider the subsets of B of points where the two lifts agree/don't agree, and show they are both open and closed.*

Lemma 5.3.5 (Paths lift). *Let $p : X \to Y$ be a covering. If $\alpha : [0, 1] \to Y$ is a path such that $\alpha(0) = y_0$ and $x_0 \in p^{-1}(y_0) \subset X$, then there exists a unique lift $\tilde{\alpha} : [0, 1] \to X$ such that $\tilde{\alpha}(0) = x_0$.*

Sketch of Proof We build the lift piece by piece, as illustrated in Figure 5.9. Choose a neighborhood U_{y_0} and corresponding V_{x_0} such that the restriction $p : V_{x_0} \to U_{y_0}$ is a homeomorphism – in particular, the restriction has an inverse, p_0^{-1}.

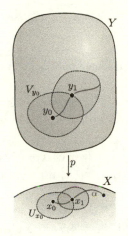

Figure 5.9 Two steps of lifting a path

Suppose that for $0 < s < s_1$ we have $\alpha(s) \in U_{y_0}$ but $\alpha(s_1) \notin U_{y_0}$. Then for $0 < s < s_1$ define the lift $\tilde{\alpha}(s) = p_0^{-1}(\alpha(s))$. Define $\tilde{\alpha}(s_1) = \lim_{s \to s_1} \tilde{\alpha}(s)$ and note that $p(\tilde{\alpha}(s_1)) = \alpha(s_1)$.

Now repeat this process using $x_1 := \alpha(s_1)$ and $y_1 := \tilde{\alpha}(s_1)$ instead of x_0, y_0 to extend the lift to $s_2 > s_1$, and we may continue until we have a full lift $\tilde{\alpha} : [0, 1] \to Y$. (Note: the compactness of $[0, 1]$ guarantees we can reach $s = 1$ in a finite number of steps.) $\qquad\square$

Using a similar strategy, one shows that homotopies of paths also lift.

Lemma 5.3.6 (Homotopies of paths lift). *Let $p : X \to Y$ be a covering and $H : [0, 1] \times [0, 1] \to Y$ a homotopy between the paths $\alpha, \beta : [0, 1] \to Y$, relative to the endpoints. Let $y_0 = \alpha(0) = H(0, 0)$ and $x_0 \in p^{-1}(y_0) \subset X$. Then there exists a lifting $\tilde{H} : [0, 1] \times [0, 1] \to X$ of H such that $\tilde{H}(0, 0) = x_0$.*

Sketch of Proof Again, the lift is built in several steps. One may start by lifting the path $\alpha = H_{|[0,1] \times \{0\}}$ as in Lemma 5.3.5. Note that at each step one may actually lift H along a basic[5] two-dimensional open set. Hence, when α is lifted, one has obtained a lift of H along some strip $[0, 1] \times [0, t_0)$. Note that $t_0 > 0$ is granted by the compactness of $[0, 1]$, which allows us to lift α in a finite number of steps. We also note that $\tilde{H}(0, t_0)$ is defined by continuity. Now we may define $\alpha_1 := H_{|[0,1] \times \{t_0\}}$ and iterate the procedure until we extend the domain of $\tilde{H}$ to all of $[0, 1] \times [0, 1]$. (Here it is the compactness

---

[5] An open set obtained as a product of two orthogonal open intervals. These are also called *box sets* and form a natural base for the product topology.

of $[0, 1] \times [0, 1]$ which allows the process to terminate in a finite number of steps.) $\qquad\qquad\square$

The next exercises explore some important consequence of Lemma 5.3.6.

Exercise 5.3.2. Let $p : X \to Y$ be a covering. If γ_1, γ_2 are homotopic loops based at y_0, and $\tilde{\gamma}_1, \tilde{\gamma}_2$ are lifts such that $\tilde{\gamma}_1(0) = \tilde{\gamma}_2(0)$, then $\tilde{\gamma}_1(1) = \tilde{\gamma}_2(1)$.

Exercise 5.3.3. Let $p : X \to Y$ be a covering and choose $y_0 \in Y$ and $x_0 \in p^{-1}(y_0) \subset X$. Show that $\pi_1(p) : \pi_1(X, x_0) \to \pi_1(Y, y_0)$ is an injective group homomorphism.

Using Lemma 5.3.5 and Lemma 5.3.6 one obtains the fundamental result about lifting arbitrary maps, which we write in the language of pointed topological spaces. Note that $A \leq B$ means A is a subgroup of B.

Proposition 5.3.7. *Let $p : (X, x_0) \to (Y, y_0)$ be a covering and $\alpha : (B, b_0) \to (Y, y_0)$ a continuous function of path-connected, pointed topological spaces. Then a lifting $\tilde{\alpha} : (B, b_0) \to (X, x_0)$ exists if and only if:*

$$\pi_1(\alpha)\,(\pi_1(B, b_0)) \leq \pi_1(p)\,(\pi_1(X, x_0)).\qquad (5.9)$$

Sketch of Proof For any $b \in B$, we choose a path γ from b_0 to b. The composition $\alpha \circ \gamma$ is a path to X and its lifting $\tilde{\gamma}$ exists by Lemma 5.3.5. We define

$$\tilde{\alpha}(b) := \tilde{\gamma}(1).$$

If the map $\tilde{\alpha}$ is well defined, then it lifts α. To show that $\tilde{\alpha}$ is well defined one uses Lemma 5.3.6 and (5.9). $\qquad\qquad\square$

A simply connected covering $q : (U, u_0) \to (Y, y_0)$ is called a **universal cover** of (Y, y_0) (sometimes one uses the name "universal cover" just to denote the topological space U). It is a bit technical to show that a universal covering always exists (see Hatcher (2002, Section 1.3)), but its uniqueness up to homeomorphism is a consequence of Proposition 5.3.7, which grants that (U, u_0) satisfies the following **universal property**.

Let $q : (U, u_0) \to (Y, y_0)$ be a universal covering for (Y, y_0). Given any covering $p : (X, x_0) \to (Y, y_0)$, there exists a unique covering $\tilde{q} : (U, u_0) \to (X, x_0)$ which lifts q. As a consequence, (U, u_0) admits a universal covering map to any covering of (Y, y_0).

Exercise 5.3.4. Show that the above universal property holds, and that it implies the uniqueness of the universal cover up to homeomorphism.

Exercise 5.3.5. Show that $\mathbb{R}$ is the universal cover of S^1, $\mathbb{R}^2$ is the universal cover for the torus, and S^2 is the universal cover for $\mathbb{P}^2(\mathbb{R})$.

We have seen that any covering of $p : (X, x_0) \to (Y, y_0)$ naturally identifies a subgroup of the fundamental group of (Y, y_0) of "loops that lifts to loops". In fact, in a fashion similar to how one constructs the universal cover, one may show that given any subgroup of $\pi_1(Y, y_0)$ there exists a corresponding covering space. The precise statement is known as the *Galois correspondence* in covering theory.

Galois correspondence in covering theory.
The function

$$\left[p : (X, x_0) \to (Y, y_0) \right] \mapsto \left[G(p) := \pi_1(p)(\pi_1(X, x_0)) \leq \pi_1(Y, y_0) \right]$$

is a bijection between the set of path-connected covers of (Y, y_0) and the set of subgroups of $\pi_1(Y, y_0)$.

Such bijection respects the poset[6] structure induced on the two sets by the covering maps and by inclusion, respectively. What this means is illustrated in the following diagram:

$$G(p_1) \leq G(p_2) \leq \pi_1(Y, y_0).$$

Remark 5.3.8. We note that by the Galois correspondence a simply connected topological space admits only trivial covers. We exploited this fact when proving the Rieman–Hurwitz formula (4.5). Also, all covers of a circle are described in Example 5.3.3. This fact will be used in the proof of the Riemann Existence Theorem (Theorem 6.2.2).

Exercise 5.3.6. Describe all connected coverings of a torus.

---

[6] Partially ordered set.

6

Counting Maps

We now introduce the counting problem for maps of Riemann Surfaces: fixing a compact Riemann Surface Y and a finite number of points $b_1, \ldots, b_n \in Y$, how many maps to Y have a specified ramification behavior over the chosen points, and are unramifed elsewhere?

Natural questions that arise are:

1. Is the number of such maps finite?
2. Does it depend on the Riemann Surface Y?
3. Does it depend on the configuration of the points b_i?

As luck would have it, the answers are about as good as possible: the number is always finite, it does depend only on the genus of Y; it also depends on the choice of ramification over the b_i but not on the position of the points. We call the answers to the question in the first paragraph *Hurwitz numbers* and we will spend the rest of this book becoming well acquainted with them.

A key reason for the favorable answers to the above questions is that maps of Riemann Surfaces are essentially "controlled by topology". We saw in Chapter 4 that maps of compact Riemann Surfaces are covering spaces away from a finite number of points of ramification: in this section we call them *ramified covers*. The *Riemann Existence Theorem* essentially says that any ramified cover corresponds to a map of Riemann Surfaces, and allows us to immediately witness that Hurwitz numbers are independent of the complex structure on Y or on the configuration of the branch points.

In subsequent chapters we will count ramified covers by analyzing the behavior of lifts of loops on Y. We conclude this chapter by looking at the simplest example: when the cover has degree 2, a loop winding around a branch point must lift to a path connecting the two inverse images of the base point. This allows us to compute all *hyperelliptic Hurwitz numbers*, counting ramified covers of degree 2 of $\mathbb{P}^1(\mathbb{C})$.

6.1 Hurwitz Numbers

We begin this section by defining the notion of isomorphism and automorphims of a map of Riemann Surfaces.

Definition 6.1.1. Two holomorphic maps of Riemann Surfaces $f : X \to Y$ and $g : \tilde{X} \to Y$ are called **isomorphic** if there is an isomorphism of Riemann Surfaces $\phi : X \to \tilde{X}$ such that $f = g \circ \phi$. An **automorphism** of $f : X \to Y$ is an isomorphism $\psi : X \to X$ such that $f = f \circ \psi$. The group of automorphisms of f is denoted $\mathrm{Aut}(f)$.

Note that if f and g are isomorphic maps via ϕ, then ϕ respects pre-images, i.e. for any $y \in Y$ the map ϕ gives a bijection $\phi : f^{-1}(y) \to g^{-1}(y)$.

Exercise 6.1.1. Consider the affine elliptic curve $E_1 = V(y^2 - (x - a_1)(x - a_2)(x - a_3)) \subset \mathbb{C}^2$ with the $a_i \in \mathbb{C}$ distinct. The projection $\pi : E_1 \to \mathbb{C}$ defined by $(x, y) \mapsto x$ is a holomorphic map. Show that the map $\sigma : E_1 \to E_1$ defined by $(x, y) \mapsto (x, -y)$ gives a nontrivial automorphism of π.

We introduce a combinatorial notion which will be used to index an important geometric (local) invariant of a map of Riemann Surfaces.

Definition 6.1.2. Let $d > 0$ be an integer. A **partition** of d is an unordered tuple of positive integers $\lambda = (k_1, k_2, \ldots)$ such that $\sum k_i = d$. Since some of the integers in the tuple may be the same, we can't think of a partition as just a set of integers. But imagine giving each part of λ a color, making sure that repeated integers get colored differently: then λ is a set of colored integers. This allows us to talk about the elements of a partition, or make sense of what a function from a partition to itself is.

The sum d of the elements of λ is called the **size** of the partition and denoted $|\lambda|$. The number of elements in λ is called the **length** of the partition and denoted $\ell(\lambda)$.

An **automorphism** of a partition λ is a bijection $\phi : \lambda \to \lambda$ such that, for every i, the equality of integers $\phi(k_i) = k_i$ holds. In simple terms one is allowed to permute the repeated values of the partition.

Although partitions are unordered collections of positive integers, it is customary to write them in non-increasing order.

Example 6.1.3. There are three distinct partitions of size 3: (3), $(2, 1)$ and $(1, 1, 1)$. Their lengths are respectively one, two and three. The last partition is the only one that admits nontrivial automorphisms. In fact, $\mathrm{Aut}(1, 1, 1) = S_3$.

Exercise 6.1.2. Write down all the partitions of 4, 5 and 6.

Definition 6.1.4. Let $f : X \to Y$ be a holomorphic map of Riemann Surfaces of degree d, let $y \in Y$ and let $f^{-1}(y) = \{x_1, \ldots, x_n\}$. Recall that for any x_i, the map f admits a local expression of the form $w = z^{k_{x_i}}$, with k_{x_i} a well-defined positive integer called the ramification index of f at x_i. We call the set $\{k_{x_1}, \ldots, k_{x_n}\}$ the **ramification profile** of f at y. Note that the ramification profile of f at y is a partition of d.

If the ramification profile of f at y is:

- $(1, \ldots, 1)$, then we say f is **unramified** over y;
- (2) or $(2, 1, \ldots, 1)$, then f has **simple ramification** over y;
- (d), where d is the degree of f, then f is **fully ramified** over y.

Example 6.1.5. For any $d > 0$, the holomorphic map $p : \mathbb{P}^1(\mathbb{C}) \to \mathbb{P}^1(\mathbb{C})$ given by the polynomial $p(x) = x^d$ is fully ramified over 0 and ∞, and unramified over every other point of $\mathbb{P}^1(\mathbb{C})$.

We are now ready to formally state the counting problem for maps of Riemann Surfaces.

Definition 6.1.6 (Hurwitz number). Let Y be a connected, compact Riemann Surface of genus g. Fix points $b_1, \ldots, b_n \in Y$ and let $\lambda_1, \ldots, \lambda_n$ be partitions of a positive integer d. We define the **Hurwitz number** as

$$H_{h \xrightarrow{d} g}(\lambda_1, \ldots, \lambda_n) = \sum_{[f]} \frac{1}{|\mathrm{Aut}(f)|}; \tag{6.1}$$

the sum in (6.1) runs over each isomorphism class of $f : X \to Y$ where

1. f is a holomorphic map of Riemann Surfaces;
2. X is connected, compact, and has genus h;
3. the branch locus of f is $B = \{b_1, \ldots, b_n\}$;
4. the ramification profile of f at b_i is λ_i.

We call a map f satisfying 1–4 a **Hurwitz cover** for the discrete data $g, h, d, \lambda_1, \ldots, \lambda_n$.

For Hurwitz covers to exist, the discrete data must satisfy the Riemann–Hurwitz formula (4.5). We note that in this case

$$\sum_{x \in X} v_x = nd - \sum_{i=1}^{n} \ell(\lambda_i). \tag{6.2}$$

Example 6.1.7. Let $Y = \mathbb{P}^1(\mathbb{C})$ and set $b_1 = 0, b_2 = \infty$. Choose $d > 0$ and let $\lambda_1 = \lambda_2 = (d)$. We compute

$$H_{0 \xrightarrow{d} 0}((d), (d)) = \frac{1}{d}. \tag{6.3}$$

Example 6.1.5 shows that $p(x) = x^d$ gives a Hurwitz cover for this discrete data. We show that any Hurwitz cover $f : X \to \mathbb{P}^1(\mathbb{C})$ is isomorphic to p, so that p is the only map we need to consider for computing the Hurwitz Number. Recall from Remark 3.2.1 that any Riemann Surface of genus 0 is isomorphic to $\mathbb{P}^1(\mathbb{C})$. Thus any Hurwitz cover is first of all a holomorphic map $f : \mathbb{P}^1(\mathbb{C}) \to \mathbb{P}^1(\mathbb{C})$ and so f is a rational function (Remark 4.5.1).

Further, f has degree d and is ramified only at r_1, r_2 where $f(r_1) = 0$, $f(r_2) = \infty$, both with ramification index d. Assuming that neither r_1 nor r_2 is ∞, we have

$$f(x) = b \frac{(x - r_1)^d}{(x - r_2)^d}$$

for some $0 \neq b \in \mathbb{C}$. To show that the map f is isomorphic to p, we must construct an isomorphism of Riemann Surfaces $\phi : \mathbb{P}^1(\mathbb{C}) \to \mathbb{P}^1(\mathbb{C})$ such that $f = p \circ \phi$. Thus ϕ is a Möbius transformation (see Exercise 4.5.2) which, in particular, sends $r_1 \mapsto 0$ and $r_2 \mapsto \infty$, i.e. we have $\phi = a(x - r_1)/(x - r_2)$ for some $0 \neq a \in \mathbb{C}$.

The equation $f = p \circ \phi$ is

$$b \frac{(x - r_1)^d}{(x - r_2)^d} = \left(a \frac{x - r_1}{x - r_2} \right)^d$$

which is satisfied by any choice of $a = b^{1/d}$. (We leave the case that one of r_1, r_2 is ∞ as an exercise.)

Thus f is isomorphic to p via ϕ, and we have only one isomorphism class of good maps to consider to compute the Hurwitz number.

Exercise 6.1.3 computes that $|\mathrm{Aut}(p)| = d$ and concludes the proof of (6.3).

Exercise 6.1.3. Let $p : \mathbb{P}^1(\mathbb{C}) \to \mathbb{P}^1(\mathbb{C})$ be the map given by $p(x) = x^d$. Show that

$$\mathrm{Aut}(p) = \{\tilde{\phi}(x) = cx | c \in \mu_d\},$$

and that this group is naturally isomorphic to μ_d, the multiplicative cyclic group of complex d-th roots of unity.

Sometimes it is useful to allow the source curves of the maps we count to be disconnected. We call the corresponding count a **disconnected Hurwitz number** and denote it by $H^{\bullet}_{\underset{h \xrightarrow{d} g}{}} (\lambda_1, \ldots, \lambda_n)$.

Before we move on too quickly, let us pause and ponder on the fact that we have not defined the genus of a disconnected Riemann Surface. If you are tempted to assume that the genus should be the sum of the genera of the connected components, remember that there is another fundamental topological invariant for surfaces, the Euler characteristic χ, which is naturally additive under disjoint unions. For a connected, compact surface

$$\chi = 2 - 2g. \tag{6.4}$$

At this point notice that if we declare the genus of a disjoint union to be the sum of the genera of the connected components, we must then appropriately modify (6.4). It is a better choice to declare (6.4) to be valid and use it to define the genus of a disconnected surface. For example, if $X = \mathbb{P}^1(\mathbb{C}) \sqcup \mathbb{P}^1(\mathbb{C})$, $\chi(X) = 2 + 2$ and therefore solving (6.4) we obtain $g(X) = -1$. The genus of a disconnected surface may be a negative integer! There are many reasons why this is the right thing to do, but one that we can readily appreciate is that with this definition the Riemann–Hurwitz formula remains unchanged when the source curve is disconnected. We conclude this discussion by unraveling the definition of genus of a disconnected curve into a purely combinatorial one.

Definition 6.1.8. Let X be a compact, orientable surface with n connected components of genera $g_1, \ldots, g_n$. Then:

$$g(X) := g_1 + \ldots + g_n + 1 - n.$$

Exercise 6.1.4. Check that using definition 6.1.8, (6.4) holds and the Riemann–Hurwitz formula remains unchanged when the source curve is disconnected.

6.2 Riemann's Existence Theorem

In Chapters 4 and 5 we saw that holomorphic functions of compact Riemann Surfaces are covering spaces away from a finite set of points. Conversely, we now show that any topological cover of a punctured Riemann Surface gives rise to a unique holomorphic map of compact Riemann Surfaces.

Definition 6.2.1. A continuous function between compact topological surfaces $p : X \to Y$ is called a **ramified cover** if there is a finite set of points $B \subset Y$ such that:

- $p^{-1}(B) \subset X$ is finite;
- $p : X \smallsetminus p^{-1}(B) \to Y \smallsetminus B$ is a covering.

We may now concisely say that maps of Riemann Surfaces are ramified covers. The following classical theorem establishes the converse statement.

Theorem 6.2.2 (Riemann's Existence Theorem). *Let Y be a compact Riemann Surface and X° a topological surface. Assume that there are a finite number of points $b_1, \ldots, b_n \in Y$ and a function $f^\circ : X^\circ \to Y \smallsetminus \{b_1, \ldots, b_n\}$ which is a topological cover of finite degree. Then there exists a unique (up to isomorphism) compact Riemann Surface X which contains X° as a dense open set (in fact X is X° plus a finite number of points) such that f° extends to $f : X \to Y$, a holomorphic map of Riemann Surfaces.*

Sketch of proof. The proof consists of two steps. The first is purely topological, and it consists in completing X° to a compact surface and extending f° to a continuous function. The second step is endowing X with a complex structure in such a way that f is a holomorphic function.

For the first step, consider one of the special points $b_1, \ldots, b_n \in Y$. To avoid unnecessary proliferation of indices, let us name the point we are considering simply b. Consider a coordinate chart φ around b and let $\Delta := \varphi^{-1}(\{|w| < 1\})$; notice that Δ is an open neighborhood of b homeomorphic to an open disk. The function $f^\circ : (f^\circ)^{-1}(\Delta \smallsetminus b) \to \Delta \smallsetminus b$ is a covering of some finite degree d by assumption. Let $U_1^\circ, \ldots, U_m^\circ$ denote the connected components of $(f^\circ)^{-1}(\Delta \smallsetminus b)$. Since $\Delta \smallsetminus b$ is homotopy equivalent to a circle, its fundamental group is $\mathbb{Z}$; analogously to the case of the circle, connected finite covers of punctured disks (such as the open sets U_i°) are themselves homeomorphic to punctured disks; further there exist positive integers $k_1, \ldots, k_m$ and homeomorphisms $\phi_i^\circ : U_i \to \{0 < |z| < 1\} \subset \mathbb{C}$ such that $\varphi \circ f^\circ \circ (\phi_i^\circ)^{-1}$ is the map $w = z^{k_i}$. For every $i = 1, \ldots, m$ now add a point x_i to X° in such a way that ϕ_i° extends to a homeomorphism $\phi_i : U_i \cup x_i \to \{|z| < 1\} \subset \mathbb{C}$, with $\phi_i(x_i) = 0$.

After repeating this process for all of the special points $b_1, \ldots, b_n \in Y$, we have added a finite number of points to X° to obtain a new topological space X, which is clearly a surface: one only needs to show that it is locally homeomorphic to a Euclidean open disk around the added points x_i and the homeomorphisms ϕ_i constructed above do precisely this. One may show that

X is compact by showing that every infinite subset of X must have a limit point, which we leave as an exercise[1]. We also note that f° extends naturally to a continuous function on all of X by sending each new point x_i to the corresponding $b \in Y$.

Finally, we must show that X may be given a complex structure in such a way that f is a holomorphic function. For every $x \in X^\circ$, one may choose an open neighborhood $U_x \subset X^\circ$ such that $f^\circ_{|U_x}$ is a homeomorphism and $f^\circ(U_x)$ is contained in some coordinate chart φ_x for Y. We then define the coordinate chart for x to be the composition $\varphi_x \circ f^\circ_{|U_x}$. With this choice, the local expression of f at x is the identity function and hence it is holomorphic.

For each of the new points x_i we use the homeomorphism ϕ_i as a local chart. As was pointed out above, a local expression for f around x_i is given by the map $w = z^{k_i}$, which is holomorphic. We leave it to the reader to check that the charts just defined are compatible, and therefore define a complex structure of X. We also leave it to the reader to verify that any other atlas on X for which f is holomorphic is in fact equivalent to the one just constructed.   □

6.3 Hyperelliptic Covers

In this section we study maps of degree 2 to the projective line. We show that, given any choice of an even number of points on $\mathbb{P}^1(\mathbb{C})$, there exists "half a map" of Riemann Surfaces of degree 2 having those points as the branch locus.

Definition 6.3.1. A Riemann Surface X is called **hyperelliptic** if it admits a holomorphic map f to $\mathbb{P}^1(\mathbb{C})$ of degree 2. Such a map f is called a **hyperelliptic cover**.

Let $f : X \to \mathbb{P}^1(\mathbb{C})$ be a hyperelliptic cover, with X a hyperelliptic Riemann Surface of genus g. Applying the Riemann–Hurwitz Formula to f gives $\sum_{x \in X} \nu_x = 2g + 2$.

Since the degree of f is 2, a point $x \in X$ is a ramification point if and only if $k_x = 2$ (i.e. if $\nu_x = 1$). Then $\sum_{x \in H} \nu_x = 2g + 2$ implies that f has $2g + 2$ distinct ramification points. It also follows that there are $2g + 2$ distinct branch points in $\mathbb{P}^1(\mathbb{C})$.

We therefore set out to understand the Hurwitz number $H_{g \to 0}^{2}((2)^{2g+2})$, where the exponent after the partition is shorthand notation to mean that we

---

[1] This is called *limit point compactness*, and it is equivalent to compactness for metrizable spaces. Surfaces are metrizable by the Urysohn metrization theorem (Munkres, 1975, page 215).

have $2g + 2$ partitions equal to (2). We show first that this number is nonzero by exhibiting explicitly a Hurwitz cover for this discrete data.

Let $b_1, \ldots, b_{2g+2}$ be fixed, distinct points in $\mathbb{P}^1(\mathbb{C})$, and we may assume without loss of generality that they are all different from ∞. Identify the affine open $\mathbb{P}^1(\mathbb{C}) \smallsetminus \infty$ with the x-axis in $\mathbb{C}^2$, and consider the affine curve X° defined by the polynomial equation

$$y^2 = \prod_{i=1}^{2g+2} (x - b_i). \tag{6.5}$$

The projection map $(x, y) \to x$ restricts to a holomorphic function $p^\circ : X^\circ \to \mathbb{C}$, which satisfies the hypothesis of the Riemann Existence Theorem. Therefore it extends uniquely to a map p from a compact Riemann Surface X to $\mathbb{P}^1(\mathbb{C})$. Each of the points b_i has only one inverse image for p, and all the b_i are branch points. Since we have an even number of b_is and the Riemann–Hurwitz formula forces $\sum_{x \in X} \nu_x$ to be even, the point ∞ is not a branch point for p. Hence the map p satisfies all the discrete data of our Hurwitz problem and it is therefore a legitimate Hurwitz cover.

Next, we must compute $\mathrm{Aut}(p)$. Let $\phi : X \to X$ be an automorphism of p and let $R \subset X$ be the ramification locus of f. If $r \in R$ then we must have $\phi(r) = r$, i.e. ϕ must fix all ramification points.

If $x \in X$ is not a ramification point, then ϕ may either fix x or switch it with the other inverse image of $p(x)$. One can argue that the subsets of $X \smallsetminus R$ consisting of points that are fixed/switched by ϕ are both open, hence one has to be empty and the other all of $X \smallsetminus R$. This leaves only two options for ϕ: either ϕ is the identity function, or it switches all non-ramification points in X. The map $\iota : X \to X$ defined as the restriction of the plane map $(x, y) \mapsto (x, -y)$ is a holomorphic map which does just that. Hence $|\mathrm{Aut}(p)| = 2$ and the contribution to $H_{g \to 0}^2((2)^{2g+2})$ by p is $1/2$.

In fact, p is the only (isomorphism class of) hyperelliptic Hurwitz cover. In order to show this, we need to analyze how paths on $\mathbb{P}^1(\mathbb{C})$ lift to a cover.

Lemma 6.3.2. *Let $f : X \to \mathbb{P}^1(\mathbb{C})$ be a hyperelliptic Hurwitz cover, with ramification locus $R = \{r_1, \ldots, r_{2g+2}\} \subset X$ and branch locus $B = \{b_1 = f(r_1), \ldots, b_{2g+2} = f(r_{2g+2})\} \subset \mathbb{P}^1(\mathbb{C})$. Let $y_0 \in \mathbb{P}^1(\mathbb{C}) \smallsetminus B$ and $f^{-1}(y_0) = \{x_1, x_2\}$.*

1. If γ is a simple loop based at y_0 which separates $\mathbb{P}^1(\mathbb{C})$ into two regions, each containing an even number of branch points, then any lift of γ via f is a loop.

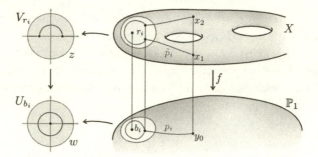

Figure 6.1 Lifting a loop winding around a branch point

2. *If γ is a simple loop based at y_0 which separates $\mathbb{P}^1(\mathbb{C})$ into two regions, each containing an odd number of branch points, then any lift of γ via f is an open path.*

Proof Since the endpoint of a lift of a path γ from a given starting point depends only on the homotopy class of γ (see Exercise 5.3.2), it suffices to show that, for every i, a loop ρ_i that winds once around b_i lifts to an open path. A loop γ that encompasses several branch points is then homotopic to a composition of the ρ_is for the corresponding branch points, and therefore its lift will close up or not according to the parity of the branch points circled.

Consider coordinate patches $U_{b_i} \ni b_i$, $V_{r_i} \ni r_i$, such that the local expression for f is $w = z^2$. Construct the loop $\rho_i = \alpha * \beta * \alpha^{-1}$ as illustrated in Figure 6.1: follow a path α from y_0 to y_1, where $y_1 \in U_{b_i}$ is the point whose local coordinate is $w = 1$. Then β is a loop based at y_1, going around the point b_i along the circle $|w| = 1$. Finally follow α backwards to y_0. Let us lift the path ρ_i starting at x_1: the endpoint of the lift of α is one of the two points with local coordinate $z = \pm 1$: without loss of generality we may assume it is $z = 1$. The lift of the loop β traces a semicircle and has endpoint $z = -1$. Finally, the lift of α^{-1} starting from $z = -1$ has to be disjoint from the lift starting from $z = 1$, and hence it must have endpoint x_2. □

We now show that p is the only map contributing to the Hurwitz number $H_{g \to 0}^{2}((2)^{2g+2})$ by showing that there is a unique ramified cover of $\mathbb{P}^1(\mathbb{C})$ branched at a specified collection of points.

Lemma 6.3.3. *Let $B = \{b_1, \ldots, b_{2g+2}\} \subset \mathbb{P}^1(\mathbb{C})$ be distinct points. There exists a unique ramified cover of $\mathbb{P}^1(\mathbb{C})$ of degree 2 with B as its branch locus.*

Proof Refer to Figure 6.2 throughout this proof.

For $i = 1, \ldots, g + 1$, let γ_i be a segment joining b_{2i-1} and b_{2i}, and assume the supports of all γ_is are disjoint.

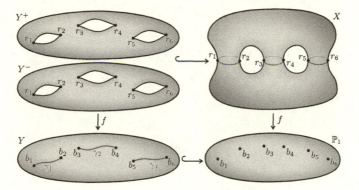

Figure 6.2 The unique degree 2 ramified cover

Consider $Y = \mathbb{P}^1(\mathbb{C}) \setminus \cup_{i=1}^{g+1} \gamma_i$, and fix a point $y_0 \in Y$. Note that Y is homeomorphic to a sphere with $g + 1$ closed discs removed.

Any ramified double cover $f : X \to \mathbb{P}^1(\mathbb{C})$ branching at B should restrict to an honest cover $f : f^{-1}(Y) \to Y$. By construction, there are no loops in Y based at y_0 that encompass an odd number of branch points. Therefore all loops in Y must lift to loops in $f^{-1}(Y)$; this forces $f^{-1}(Y) \to Y$ to be a trivial cover: $f^{-1}(Y) \cong Y^+ \cup Y^-$ is homeomorphic to two disjoint copies of Y.

Finally, we observe that for each of the γ_i, $f^{-1}(\gamma_i)$ is a circle with distinguished points r_{2i-1}, r_{2i}. The ramified cover X is then obtained uniquely by gluing the boundaries of the removed discs to the circles as shown in Figure 6.2. $\qquad\square$

Since we have shown that there is a unique genus g hyperelliptic Hurwitz cover, we now understand all hyperelliptic Hurwitz numbers. Recalling that a hyperelliptic cover admits one nontrivial automorphism, we have

$$H_{g \xrightarrow{2} 0}((2)^{2g+2}) = \frac{1}{2}.$$

7

Counting Monodromy Representations

At the end of Chapter 6 we introduced an interesting approach to the study of a hyperelliptic cover $f : X \to \mathbb{P}^1(\mathbb{C})$: by performing appropriate cuts on both source and target Riemann Surfaces, we reduced the space upstairs to be just two disjoint copies of the space downstairs. We then reconstructed X by observing that there is a unique (orientation-preserving) way to reglue the two copies along their boundaries in a way compatible with the existence of a global map f.

In this chapter we wish to generalize this idea to what might be called the *Ikea approach* to covers. Suppose you ordered online your favorite cover of Riemann Surfaces $f : X \to Y$. For the sake of saving on shipping costs, the warehouse would like to cut Y along appropriate segments in such a way that it could be flattened to a topological disk P. If the branch locus B of f is contained in the cuts, then cutting X along the inverse image of the cuts in Y produces d disjoint identical copies of P, which with Scandinavian precision would be labeled $P_1, \ldots, P_d$. What you will get in the mail is an envelope containing these $d + 1$ disks and, hopefully, a manual of assembly instructions.

A way to provide assembly instructions is to specify, for every loop on Y, its lifts to X. For example, suppose that a loop ρ exits P at a point x and re-enters it at another point y (think of P as some kind of *Pac-Man* screen governed by the geometry of Y). Suppose you are also told that when you lift ρ starting from the polygon P_1 you end up in polygon P_3. This information tells you that you should glue the points x and y together; and that you should glue the point corresponding to x in P_1 to the point corresponding to y in P_3. It's easy to imagine that if you know such information for every possible loop, you could eventually glue back all sides of P to reconstruct Y and all sides of the various P_is to obtain X. While at first this seems like a daunting amount of information to control, because the endpoints of lifts of loops are invariant under homotopy, all such information is contained in a group homomorphism $\Phi : \pi_1(Y \smallsetminus B, y_0) \to S_d$, called the *monodromy representation* of the cover f.

90

In this chapter we become familiar with monodromy representations. We develop a very tight dictionary between ramified covers and monodromy representations, which allows us to translate the Hurwitz problem of counting maps of Riemann Surfaces into a group theoretic problem of counting appropriate monodromy representations. This translation is extremely helpful. For example, because fundamental groups of punctured surfaces are finitely generated and S_d is a finite group, we see immediately that the answer to the Hurwitz counting problem is always a finite number.

7.1 From Maps to Monodromy

Let $f : X \to Y$ be a degree d holomorphic map of connected Riemann Surfaces with branch locus $B = \{b_1, \ldots, b_n\} \subset Y$. Choose a $y_0 \notin B$ and consider a loop $\gamma : [0, 1] \to Y \smallsetminus B$ based at y_0 as shown in Figure 7.1. Choosing a preimage $x \in f^{-1}(y_0)$, γ lifts to a path $\tilde{\gamma}_x$ in X starting at x. Since $\gamma(1) = y_0$, the endpoint of $\tilde{\gamma}_x$ is a preimage of y_0 (possibly different from x). We can thus associate to γ a function

$$\sigma_\gamma : f^{-1}(y_0) \to f^{-1}(y_0)$$

defined by $\sigma_\gamma(x) = \tilde{\gamma}_x(1)$.

Exercise 7.1.1. Show that the function σ_γ is a bijection.

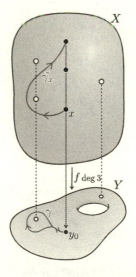

Figure 7.1 Lifting a generic loop

Exercise 7.1.1 shows that the loop γ yields a permutation σ_γ of the preimages of y_0. We may write σ_γ as an element of the symmetric group S_d by labeling the d preimages of y_0 with the numbers $1, 2, \ldots, d$ (see Figure 7.2).

Definition 7.1.1. A y_0-**labeled map** is a pair (f, L), where $f : X \to Y$ is a degree d map of Riemann Surfaces and $L : f^{-1}(y_0) \to \{1, \ldots, d\}$ is a bijection. Note that this forces y_0 to not be a branch point for f. Then L is called a **labeling** of the inverse images of y_0.

An **isomorphism** of y_0-labeled maps $(f_1, L_1), (f_2, L_2)$ consists of an isomorphism of Riemann Surfaces $\phi : X_1 \to X_2$ such that

$$f_2 \circ \phi = f_1 \qquad L_2 \circ \phi = L_1.$$

Remark 7.1.2. Choosing a labeling of the inverse images of $f^{-1}(y_0)$ to associate a permutation of numbers to σ_γ is an analogous procedure to choosing a basis in a vector space to assign coordinates to a given vector. The following exercise should then be reminiscent of the linear algebra concept of "changing basis".

Exercise 7.1.2. Show that the two elements $\sigma_1, \sigma_2 \in S_d$ associated to σ_γ through two distinct labelings of $f^{-1}(y_0)$ are conjugate to each other, i.e.

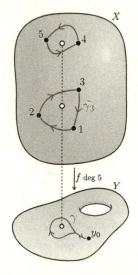

Figure 7.2 Covering map with lifted paths giving the permutation (123)(45) of the labeled preimages

there exists $\omega \in S_d$ such that $\sigma_2 = \omega\sigma_1\omega^{-1}$. Describe the relation between the permutation ω and the two labelings of $f^{-1}(y_0)$.

Example 7.1.3. Let $f : \mathbb{C} \to \mathbb{C}$ be defined by $f(x) = x^4$. Choose $y_0 = 1$ and consider the loop $\gamma(t) = \exp(2\pi t i)$. The inverse image of 1 is $f^{-1}(1) = \{1, i, -1, -i\}$, and the lifts of γ

$$\tilde{\gamma}_1(t) = \exp\left(\frac{\pi}{2}ti\right), \qquad \tilde{\gamma}_i(t) = \exp\left(\frac{\pi}{2}(ti + 1)\right)$$

$$\tilde{\gamma}_{-1}(t) = \exp\left(\frac{\pi}{2}(ti + 2)\right), \qquad \tilde{\gamma}_{-i}(t) = \exp\left(\frac{\pi}{2}(ti + 3)\right),$$

give rise to the permutation

$$\sigma_\gamma : 1 \mapsto i \mapsto -1 \mapsto -i \mapsto 1.$$

For $k = 1, \ldots, 4$, choosing the labeling $i^k \leftrightarrow k$ then gives the cycle $\sigma_\gamma = (1234) \in S_d$.

By Exercise 5.3.2, the permutation σ_γ only depends on the homotopy class of the loop γ; the permutation associated to the concatenation of two loops is the composition of the associated permutations:

$$\sigma_{\gamma*\eta} = \sigma_\eta \circ \sigma_\gamma. \tag{7.1}$$

Remark 7.1.4. Remember that multiplication in S_d is composition of functions; we emphasize this in (7.1) by using the symbol $\circ$. We adopt the common convention to drop the symbol and to write a product of permutations from right to left.

Thus a y_0-labeled map $(f : X \to Y, L)$ gives a group homomorphism

$$\Phi : \pi_1(Y \setminus B, y_0) \to S_d$$

defined by $\Phi : \gamma \mapsto \sigma_\gamma$. These group homomorphisms, called **monodromy representations,** play a central role in our story; we now wish to encode data about the ramification profile over each of the branch points: to do so, we begin by recalling a basic notion about the symmetric group S_d.

Definition 7.1.5. A permutation whose cycle decomposition consists of disjoint cycles of length $\{l_1, \ldots, l_k\}$ is said to have **cycle type** $\{l_1, \ldots, l_k\}$.

For example, the permutation $(12)(45)(368) \in S_8$ has cycle type $\{3, 2, 2, 1\}$. The length-one cycle (7) is omitted from the cycle notation for the permutation, but it is remembered in the cycle type. This way, the cycle type of a permutation in S_d is always a partition of d.

Exercise 7.1.3.

1. Show that if $\tau = (a_1, a_2, \ldots, a_n) \in S_d$ is a cycle, then for any $\sigma \in S_d$ we have $\sigma \tau \sigma^{-1} = (\sigma(a_1), \sigma(a_2), \ldots, \sigma(a_n))$.
2. Show that two permutations in S_d are conjugate if and only if they have the same cycle type. (This shows that conjugacy classes in S_d may be indexed by partitions of d.)

Let us go back to a y_0-labeled map $f : X \to Y$ yielding the group homomorphism Φ. If f has ramification profile $\lambda = \{k_1, \ldots, k_l\}$ at a branch point $b \in Y$ and ρ is the class of a simple loop winding once around b, then the permutation $\Phi(\rho) = \sigma_\rho \in S_d$ has cycle type λ: For $j = 1, \ldots, l$, let r_j be an inverse image of b and choose local charts around r_j and b in such a way that the local expression for f is $w = z^{k_j}$. Note $\rho \sim \alpha * \beta * \alpha^{-1}$, with β a simple parameterization of $\{|w| = 1\}$ and α a path connecting y_0 to the point with local coordinate $w = 1$. Then, as in Example 7.1.3, σ_ρ cyclically permutes the k_j roots of unity $\{z^{k_j} = 1\}$. Since this happens around each of the ramification points above b, σ_ρ consists of l disjoint cycles each of length k_j.

Definition 7.1.6 (Monodromy Representation). Let Y be a connected Riemann Surface of genus g let and $y_0, b_1, \ldots, b_n \in Y$. Let $\lambda_1, \ldots, \lambda_n$ be partitions of a positive integer d.

A **monodromy representation** of type $(g, d, \lambda_1, \ldots, \lambda_n)$ is a group homomorphism $\Phi : \pi_1(Y \smallsetminus \{b_1, \ldots, b_n\}, y_0) \to S_d$ such that, if ρ_k is the homotopy class of a small loop around b_k, the permutation $\Phi(\rho_k)$ has cycle type λ_k.

If in addition the subgroup $\text{Im } \Phi \leq S_d$ acts transitively on the set $\{1, 2, \ldots, d\}$, we say Φ is a **connected monodromy representation**.

Remark 7.1.7. Notice that a monodromy representation is a purely topological construction: it doesn't know about the complex structure on Y or on the position of the punctures $b_1, \ldots, b_n$.

Hence we have seen that a degree d y_0-labeled map of Riemann Surfaces $f : X \to Y$ such that the ramification profile at each branch point is given by λ_i gives rise to a monodromy representation Φ of type $(g_Y, d, \lambda_1, \ldots, \lambda_n)$.

Exercise 7.1.4. Show that two isomorphic y_0-labeled maps give rise to the same monodromy representation:

$$(f_1, L_1) \cong (f_2, L_2) \implies \Phi_1 = \Phi_2.$$

Example 7.1.8. We describe monodromy representations for $Y = \mathbb{P}^1(\mathbb{C})$. Choose a finite subset $B = \{b_1, \ldots, b_n\} \subset \mathbb{P}^1(\mathbb{C})$. The punctured sphere $\mathbb{P}^1(\mathbb{C}) \smallsetminus B$ is homotopic a point with $n-1$ loops attached to it (see Figure 7.3).

The fundamental group of this space is the free group $\mathbb{F}_{n-1}$ with $n-1$ generators $\rho_1, \ldots, \rho_{n-1}$, representing loops that wind around each of the first $n-1$ branch points. Thus, for a chosen d, a group homomorphism $\Phi : \pi_1(\mathbb{P}^1(\mathbb{C}) \smallsetminus B, y_0) \to S_d$ is given by a choice of images $\Phi(\rho_k) \in S_d$ with no restrictions.

A more symmetric presentation of the fundamental group of a punctured sphere chooses small loops going around all of the branch points as generators, and realizes $\pi_1(\mathbb{P}^1(\mathbb{C}) \smallsetminus B, y_0)$ as a quotient of $\mathbb{F}_n$ by the relation $\rho_1 \ldots \rho_n = e$. This corresponds to the geometric fact that the loop $\rho_1 * \cdots * \rho_n$ is homotopic to a loop around all the b_k, which can then contract to a point along the opposite side of the sphere. Thus a monodromy representation is given by a choice of n-elements in S_d subject to the relation $\Phi(\rho_n) \circ \cdots \circ \Phi(\rho_1) = e \in S_d$.

The following exercises explain the terminology "connected monodromy representation".

Exercise 7.1.5. Let $Y = \mathbb{P}^1(\mathbb{C})$ and X be two disjoint copies of $\mathbb{P}^1(\mathbb{C})$. Define $f : X \to Y$ where both copies of $\mathbb{P}^1(\mathbb{C})$ in X map to Y via x^3. Set $y_0 = 1$. Label the preimages $f^{-1}(y_0)$ and show that the associated monodromy representation is *not* a connected monodromy representation.

Exercise 7.1.6. Let $f : X \to Y$ be a degree d holomorphic map of Riemann Surfaces. Show that, for any choice of base point y_0 and of labelings of

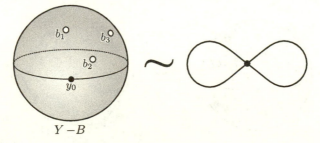

Figure 7.3 Punctured sphere and rose graph

$f^{-1}(y_0)$, the corresponding monodromy representation Φ is connected if and only if X is a connected Riemann Surface.

7.2 From Monodromy to Maps

In this section we show how a monodromy representation can be used to construct a topological cover of a punctured Riemann Surface; then by Riemann's Existence Theorem this cover can be completed to a holomorphic map of Riemann Surfaces.

The idea to have in mind is the following: given a map of Riemann Surfaces, one may choose an appropriate open dense subset of the base curve, over which the map is a trivial covering. The source curve may then be reconstructed by appropriately gluing together the boundaries of the various disconnected pieces. The monodromy representation associated to f gives "assembly instructions" to do just that. We first illustrate this procedure in a concrete example.

Example 7.2.1. Let $Y = \mathbb{P}^1(\mathbb{C})$ and choose $y_0, b_1, b_2, b_3 \in \mathbb{P}^1(\mathbb{C})$ as in Figure 7.4. Let ρ_k be a small loop around b_k and define a (connected) monodromy representation $\Phi : \pi_1(Y \smallsetminus \{b_1, b_2, b_3\}, y_0) \to S_3$ by

$$\rho_1 \mapsto (123), \quad \rho_2 \mapsto (13), \quad \rho_3 \mapsto (12).$$

Choose a point $p \in \mathbb{P}^1(\mathbb{C})$ and draw segments from p to each of the b_k. Then $Y \smallsetminus \{b_1, b_2, b_3\}$ is homeomorphic to the identification polygon P in Figure 7.4.

To construct a cover whose associated monodromy is Φ, we take three copies of the polygon P (which we label P_1, P_2, P_3) mapping to P via the identity function. We use the permutations $\Phi(\rho_k)$ to indicate how these polygons should be glued to each other. Refer to Figure 7.5 for this discussion.

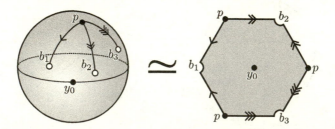

Figure 7.4 Sliced $\mathbb{P}^1(\mathbb{C})$ and homeomorphic identification polygon

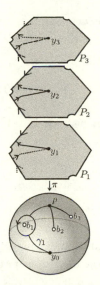

Figure 7.5 Sliced $\mathbb{P}^1(\mathbb{C})$ and the lift of ρ_1, one of the three fundamental loops ρ_1, ρ_2, ρ_3 generating $\pi_1(\mathbb{P}^1(\mathbb{C}) \smallsetminus \{b_1, b_2, b_3\})$

Label y_1, y_2, y_3 the points which correspond to y_0 in each respective polygon P_k and consider lifting the loop ρ_1 starting at y_1. The lift exits P_1 through the top-left side, and in order to have $\Phi(\rho_1) : 1 \mapsto 2$ the lift must enter P_2 on the bottom-left side and end at y_2. This gives us the "gluing instruction" of identifying the top-left side of P_1 with the bottom-left side of P_2.

Lifting ρ_1 again starting at y_2 yields a path which ends at y_3 (since $\Phi(\rho_1)$: $2 \mapsto 3$) and gives a similar identification of sides for P_2 and P_3. Lifting ρ_1 once more, starting at y_3, brings us back to y_1 and gives an identification of sides between P_3 and P_1.

We repeat this process with the loops ρ_2, ρ_3 and obtain pairwise identifications of all sides of $P_1 \cup P_2 \cup P_3$. We leave it as an exercise that such a (disconnected) identification polygon corresponds to a sphere with five punctures.

We have obtained a topological cover $\pi^\circ : X^\circ := (\cup P_k / \sim) \to Y \smallsetminus \{b_1, b_2, b_3\}$ and by Riemann's Existence Theorem (Theorem 6.2.2) there is a unique map of compact Riemann Surfaces $\pi : X \to Y$ which extends π°.

We now generalize this construction to show that any monodromy representation comes from a labeled map of Riemann Surfaces.

Theorem 7.2.2. *Given a monodromy representation* Φ *of type* $(g, d, \lambda_1, \ldots, \lambda_n)$, *for any Riemann Surface* Y *of genus* g, *and* $B = \{b_1, \ldots, b_n\} \in Y$ *there*

exists a y_0-labeled map of Riemann Surfaces covering Y with branch locus B,
whose associated monodromy representation is Φ. Such a map is unique up to
isomorphism of y_0-labeled maps.

Proof Construct a graph Γ on Y as follows: represent Y as an identification
polygon of type $\alpha_1 \beta_1 \bar{\alpha}_1 \bar{\beta}_1 \cdots \alpha_g \beta_g \bar{\alpha}_g \bar{\beta}_g$; Γ consists of the boundary of the
polygon together with n segments s_j from one of the vertices of the polygon
to each of the branch points. By "opening up" each of the s_j one may view Γ
as the boundary of the identification polygon

$$P := s_1 \bar{s}_1 \cdots s_n \bar{s}_n \alpha_1 \beta_1 \bar{\alpha}_1 \bar{\beta}_1 \cdots \alpha_g \beta_g \bar{\alpha}_g \bar{\beta}_g, \qquad (7.2)$$

which shows in particular that $Y \smallsetminus \Gamma$ is homeomorphic to a disc. This
construction is illustrated in Figure 7.6.

Consider d copies of the polygon P denoted $P_1, \ldots, P_d$, each mapping to
P via the identity function. The inverse images of y_0 are denoted $y_1, \ldots, y_d$.
The sides of the polygon P_k are denoted by $\alpha_{i,k}, \beta_{i,k}, s_{j,k}$.

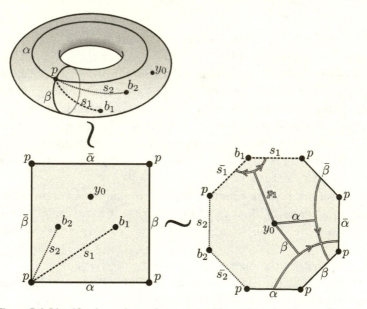

Figure 7.6 Identification polygon for a torus with two branch points, and generators of the fundamental group of the punctured surface. Note that the loops α
and β are essentially the same as the sides of the identification polygon labeled
correspondingly. To make this statement true on the nose, we should draw a path
connecting y_0 to a vertex of the polygon and conjugate the side of the polygon
with it, so as to make the side of the polygon a loop based at y_0. We don't draw
such a path to avoid cluttering the picture.

We now construct a topological surface and a ramified cover to Y by giving appropriate identifications to the sides of $P_1, \ldots, P_d$. The necessary information is contained in the monodromy representation Φ. The fundamental group $\pi_1(Y \smallsetminus B, y_0)$ is generated by loops $\alpha_i, \beta_i, \rho_j$: the α_i and β_i loops parameterize the sides of P labeled correspondingly. The loop ρ_i winds around the i-th branch point.

Observe that the loop ρ_i "exits" the polygon P at the side denoted s_i and "re-enters" at $\bar{s}_i$. The loop α_i (respectively β_i) exits at the side labeled β_i (respectively $\bar{\alpha}_i$) and re-enters at $\bar{\beta}_i$ (respectively α_i).

When ρ_j is lifted starting at y_k, it exits the polygon P_k at $s_{j,k}$ and enters the polygon labeled by the image of k in the permutation $\Phi(\rho_j)$. Therefore we are given the identification of the sides

$$s_{j,k} \sim \bar{s}_{j,\Phi(\rho_j)(k)}.$$

Proceeding similarly with the lifts of the α and β loops, we construct a (disconnected) identification polygon

$$P_1 \cup P_2 \cup \ldots \cup P_d / \sim$$

where, for $i = 1, \ldots, g$, $j = 1, \ldots, n$, $k = 1, \ldots, d$ the identifications $\sim$ are given by:

$$
\begin{aligned}
s_{j,k} &\sim \bar{s}_{j,\Phi(\rho_j)(k)}, \\
\bar{\alpha}_{i,k} &\sim \alpha_{i,\Phi(\beta_i)(k)}, \\
\beta_{i,k} &\sim \bar{\beta}_{i,\Phi(\alpha_i)(k)}.
\end{aligned}
\tag{7.3}
$$

The map $Id_1 \cup \ldots \cup Id_d$ descends to the quotient to define a function which is, by construction, a ramified cover of Y branched at B. By Riemann's Existence Theorem there is a unique complex structure on X that makes f differentiable.

Now assume that $f : X \to Y$ is a y_0 labeled map of Riemann Surfaces which also gives rise to the monodromy representation Φ. The restriction of f to $X \smallsetminus f^{-1}(\Gamma)$ is a covering of a simply connected set, and hence it is homeomorphic to d disjoint discs, with f restricting on each of them to a homeomorphism with $Y \smallsetminus \Gamma$.

Define a map $\varphi : X \smallsetminus f^{-1}(\Gamma) \to P_1 \cup P_2 \cup \ldots \cup P_d$ as follows: if x is in the connected component of $X \smallsetminus f^{-1}(\Gamma)$ containing the point marked i, then $\varphi(x)$ is defined to be the point $f(x)$ in the polygon P_i. The map so defined is a homeomorphism between $X \smallsetminus f^{-1}(\Gamma)$ and the union of the interiors of the polygons P_i. The fact that f gives the monodromy representation Φ implies that φ extends to a homeomorphism $\bar{\varphi}$ which yields the isomorphism of y_0 labeled maps:

$$\overline{\varphi} : X \xrightarrow{\quad \cong \quad} P_1 \cup P_2 \cup \ldots \cup P_d / \sim \qquad (7.4)$$

$$f \searrow \qquad \nearrow Id_1 \cup \ldots \cup Id_d$$

$$Y$$

$\square$

7.3 Monodromy Representations and Hurwitz Numbers

Let Y be a Riemann Surface of genus g. In Sections 7.1 and 7.2 we constructed inverse bijections between the following two sets:

$$\left\{ \begin{array}{c} \text{Isomorphism classes of } (f : X \to Y, L) \\ y_0\text{-labeled maps of Riemann Surfaces,} \\ X \text{ a connected Riemann Surface,} \\ f \text{ branches over } B = \{b_1, \ldots, b_n\} \\ \text{with ramification profile } \lambda_i \text{ over } b_i \end{array} \right\} \leftrightarrow \left\{ \begin{array}{c} \Phi : \pi_1(Y \setminus B, y_0) \to S_d \\ \text{connected monodromy} \\ \text{representations of type} \\ (g, d, \lambda_1, \ldots, \lambda_n) \end{array} \right\}$$

$$(7.5)$$

The set on the left-hand side of (7.5) is almost what we want to count. Theorem 7.3.1 gives the precise relationship between Hurwitz numbers and connected monodromy representations.

Theorem 7.3.1. *Let M be the set of connected monodromy representations of type $(g, d, \lambda_1, \ldots, \lambda_n)$. Then*

$$H_{h \xrightarrow{d} g}(\lambda_1, \ldots, \lambda_n) = \frac{|M|}{d!}$$

where h is determined by the Riemann–Hurwitz Formula.

Proof To prove Theorem 7.3.1, we must count the number of distinct isomorphism classes of y_0-labeled maps for a given Hurwitz cover f. Given a map $f : X \to Y$ and a smooth point $y_0 \in Y$, there are $d!$ ways to label the preimages $f^{-1}(y_0)$.

An automorphism of f produces an isomorphism of y_0-labeled maps, where the map f is held constant but the labeling changes. Specifically, if L is a labeling of the elements of $f^{-1}(y_0)$ and $\varphi \in \mathrm{Aut}(f)$, then we define the associated labeling $\varphi \bullet L := L \circ \varphi^{-1}$, and we have that $(f, L) \cong (f, \varphi \bullet L)$. Exercise 7.3.2 shows that this is a free left group action of $\mathrm{Aut}(f)$ on the set of labelings of $f^{-1}(y_0)$. Hence the number of isomorphism classes of y_0-labeled maps for the given map f, which is also the number of distinct monodromy representations arising from f by different labelings of $f^{-1}(y_0)$, is

$$m_f = d!/|\text{Aut}(f)|.$$

Finally, we have

$$H_{h \xrightarrow{d} g}(\lambda_1, \ldots, \lambda_n) = \sum_{[f]} \frac{1}{|\text{Aut}(f)|} = \sum_{[f]} \frac{m_f}{d!} = \frac{1}{d!} \sum_{[f]} m_f = \frac{1}{d!}|M|.$$

$$(7.6)$$

$\square$

Exercise 7.3.1. Use Theorem 7.3.1 to compute $H_{0 \to 0}^d((d), (d))$ for $d > 0$. Compare your answer with Example 6.1.7.

Exercise 7.3.2. Let $f : X \to Y$ be a degree d map of connected Riemann Surfaces, and choose $y_0 \in Y$ a smooth point.

1. Show that the action $\varphi \bullet L = L \circ \varphi^{-1}$ for $\varphi \in \text{Aut}(f)$ and $L : f^{-1}(y_0) \to \{1, 2, \ldots, d\}$ a labeling is a *left group action* on the set of labelings for the map f.
2. Show that the action is *free*, i.e. that $\varphi \bullet L = L$ implies $\varphi = I_X$, the identity map.

   The following exercise is meant to demonstrate in a concrete example the relationship between automorphisms, labelings and monodromy representations that we just exploited in general in the proof of Theorem 7.3.1.

Exercise 7.3.3. Consider the map $p : \mathbb{P}^1(\mathbb{C}) \to \mathbb{P}^1(\mathbb{C})$ given by $p(x) = x^3$, and choose $y_0 = 1$. Label the preimages of y_0 in all possible ways. Observe that only two distinct monodromy representations are produced by these labelings. Note that $\text{Aut}(p) = \mu_3$ acts freely on the set of labelings and the orbit of this action are precisely the labelings that yield the same monodromy representation.

   The transitivity requirement in the definition of connected monodromy representations insures that the associated Riemann Surface constructed is connected. By not requiring the monodromy representations to be connected, we obtain a theorem analogous to Theorem 7.3.1 relating monodromy representations to disconnected Hurwitz covers.

Theorem 7.3.2. *Let $M^\bullet$ be the set of (not necessarily connected) monodromy representations of type $(g, d, \lambda_1, \ldots, \lambda_n)$. Then we have*

$$H^{\bullet}_{\underset{h \xrightarrow{d} g}{}}(\lambda_1, \ldots, \lambda_n) = \frac{|M^{\bullet}|}{d!}. \tag{7.7}$$

We know from Exercise 5.2.4 that the fundamental group of a genus g surface with n punctures can be presented with $2g + n$ generators. Since S_d is a finite group, the set of monodromy representations of type $(g, d, \lambda_1, \ldots, \lambda_n)$ is a (proper) subset of the set $(S_d)^{2g+n}$, corresponding to all possible choices for the images of the generators. Hence we have this immediate corollary of Theorems 7.3.1 and 7.3.2.

Corollary 7.3.3. *For every integer g, positive integer d and $\lambda_1, \ldots, \lambda_n$ partitions of d, the Hurwtiz numbers $H_{\underset{h \xrightarrow{d} g}{}}(\lambda_1, \ldots, \lambda_n)$ and $H^{\bullet}_{\underset{h \xrightarrow{d} g}{}}(\lambda_1, \ldots, \lambda_n)$ are finite.*

7.4 Examples and Computations

Theorems 7.3.1 and 7.3.2 allow us to compute many Hurwitz numbers that are out of reach when only considering the geometric definition.

Example 7.4.1 (Hyperelliptic Hurwitz Numbers). We begin by revisiting the computation of hyperelliptic Hurwitz numbers. Recall from Section 6.3 that

$$H_{\underset{g \xrightarrow{2} 0}{}}((2)^{2g+2}) = H^{\bullet}_{\underset{g \xrightarrow{2} 0}{}}((2)^{2g+2}) = \frac{1}{2}. \tag{7.8}$$

We observe that, since we have at least one point with full ramification, all Hurwitz covers are connected. Correspondingly, monodromy representations $\Phi : \pi_1(Y \setminus B, y_0) \to S_2$ are always connected.

Recalling that $\pi_1(Y \setminus B, y_0)$ can be presented by $2g + 2$ generators ρ_i subject to the relation $\rho_1 \cdots \rho_{2g+2} = e$, a monodromy representation of type $(g = 0, d = 2, (2)^{2g+2})$ consists of the choice of $2g + 2$ transpositions in S_2 such that their product is the identity. There is only one 2-cycle in S_2 and its even power is the identity, thus there is exactly one monodromy representation of type $(0, 2, (2)^{2g+2})$. Dividing by 2!, we obtain (7.8).

Example 7.4.2. We compute $H_{\underset{0 \xrightarrow{3} 0}{}}((3), (2, 1)^2)$. Note that the ramification index (3) implies that the image of any monodromy representation Φ must contain a three-cycle and so acts transitively on $\{1, 2, 3\}$. Thus all monodromy representations in question are connected.

Any monodromy representation must have $\rho_1 \mapsto (123)$ or (132), $\rho_2 \mapsto (12)$ or (13) or (23), and $\Phi(\rho_3)$ uniquely determined as $\Phi(\rho_3) = \Phi(\rho_1)^{-1} \circ$

$\Phi(\rho_2)^{-1}$. For any choice of a three-cycle and a transposition, their product gives an odd element in S_3, and hence automatically a transposition. Thus there are exactly $2 \cdot 3 \cdot 1 = 6$ choices for the images of the ρ_i. Dividing by 3!, we obtain $H_{0 \xrightarrow{3} 0,3}((3), (2,1)^2) = 1$.

The fact that transpositions are the only odd elements in S_3 is extremely convenient for the computation of degree 3 Hurwitz numbers with at least one (and hence at least two) simple branch points: letting the simple branch points be the last, one may choose freely images for all ρ_is except the last, which is then automatically determined as a transposition. Let us witness in the following exercise how things get (a little) more complicated when the cycle type of the (image of the) last generator is not uniquely determined.

Exercise 7.4.1. Compute $H_{0 \xrightarrow{3} 0}((3), (2,1)^2)$ again, but this time leave the image of ρ_1 until last, letting it be determined by the choice of the two-cycles.

Example 7.4.3. We now compute our first example where there is a difference between connected and disconnected Hurwitz numbers:

$$H^{\bullet}_{0 \xrightarrow{3} 0}((2,1)^4) = \frac{9}{2} \qquad H_{0 \xrightarrow{3} 0}((2,1)^4) = 4. \tag{7.9}$$

To count monodromy representation $\Phi : \pi_1(Y \smallsetminus \{b_1, \ldots, b_4\}) \to S_3$, we may choose freely any three transpositions for the images of ρ_1, ρ_2, ρ_3; then the image of ρ_4 is uniquely determined and a two-cycle. This gives $3 \cdot 3 \cdot 3 \cdot 1$ choices, and dividing by 3! we obtain 9/2.

Not all of these monodromy representations Φ are connected – it is possible to have the image of Φ not act transitively on $\{1, 2, 3\}$. This happens precisely when each of $\rho_1, \ldots, \rho_4$ is sent to the same transposition (this gives a valid homomorphism since the fourth power of a transposition is the identity). Thus there are three disconnected monodromy representations, and subtracting them off gives $(3^3 - 3)/3! = 4$.

The difference of 1/2 between the connected and disconnected Hurwitz numbers comes from the disconnected cover consisting of an elliptic curve mapping to the line (i.e. $\mathbb{P}^1(\mathbb{C})$) as a double cover and a line mapping isomorphically.

Exercise 7.4.2. In this exercise we complete the computation of base genus 0 Hurwitz numbers in degree 3. For $m > 0$, $n > 0$ prove that:

$$H^{\bullet}_{h \xrightarrow{3} 0}((3)^m, (2,1)^{2n}) = H_{h \xrightarrow{3} 0}((3)^m, (2,1)^{2n}) = 2^{m-1} 3^{2n-2}, \tag{7.10}$$

with $h = m + n - 2$.

The case when all ramification is simple requires us to study separately the connected and disconnected cases:

$$H^\bullet_{h\xrightarrow{3}0}((2,1)^{2n}) = \frac{3^{2n-2}}{2} \qquad H_{h\xrightarrow{3}0}((2,1)^{2n}) = \frac{3^{2n-2}}{2} - \frac{1}{2}. \qquad (7.11)$$

The most delicate case is when all ramification is of type (3). In this case prove that:

$$H^\bullet_{h\xrightarrow{3}0}((3)^m) = H_{h\xrightarrow{3}0}((3)^m) = \frac{1}{9}(2^{m-1} + (-1)^m). \qquad (7.12)$$

Even when the discrete data satisfies the Riemann–Hurwitz formula, it is possible that there are no Hurwitz covers for that data. Remarkably, finding necessary and sufficient conditions for a Hurwitz number to be nonzero is an open problem, known as the *Hurwitz existence problem* (see a discussion, for example, in Caporaso (2014, Section 2.2)). The simplest example of this phenomenon is illustrated in this exercise.

Exercise 7.4.3. Compute the following Hurwitz number:

$$H_{0\xrightarrow{4}0}((3),(2,2)^2) = 0.$$

We conclude this section with some examples of Hurwitz numbers for a base curve of genus higher than 0.

Example 7.4.4. Let us compute $H^\bullet_{1\xrightarrow{d}1}(\varnothing)$. Hurwitz covers for this discrete data are unramified, and have a complex torus T as domain. The fundamental group $\pi_1(T, y_0) \cong \mathbb{Z} \oplus \mathbb{Z}$ may be presented with two generators, α, β, subject to the relation $\alpha\beta\alpha^{-1}\beta^{-1} = e$, i.e. $\langle \alpha, \beta | \alpha\beta\alpha^{-1}\beta^{-1} = e \rangle$.

Hence any homomorphism $\Phi : \pi_1(T) \to S_d$ is determined by a choice of $\Phi(\alpha)$, $\Phi(\beta) \in S_d$ (call these images σ_1, σ_2 respectively) such that $\sigma_1\sigma_2\sigma_1^{-1}\sigma_2^{-1} = e$, i.e. $\sigma_1\sigma_2 = \sigma_2\sigma_1$. Thus, after σ_1 is chosen, σ_2 must be chosen from the centralizer of σ_1. Recall that the **centralizer** of an element $g \in G$ is the subgroup $\xi(g) = \{h \in G | hg = gh\}$ of all elements which commute with g.

The number of monodromy representations of type $(1, d, \varnothing)$ may be expressed as $\sum_{\sigma_1 \in S_d} |\xi(\sigma_1)|$. Theorem 7.3.1 yields $H^\bullet_{1\xrightarrow{d}1}(\varnothing) = (1/d!) \sum_{\sigma_1 \in S_d} |\xi(\sigma_1)|$. The Orbit-Stabilizer Theorem (applied to a group G acting on itself by conjugation) implies $|G| = |\xi(g)||C_g|$ for any $g \in G$ and where C_g is the conjugacy class of g. In this case this gives $d! = |\xi(\sigma_1)||C_{\sigma_1}|$ for any $\sigma_1 \in S_d$, and hence

$$H^{\bullet}_{1 \xrightarrow{d} 1}(\varnothing) = \frac{1}{d!} \sum_{\sigma_1 \in S_d} \frac{d!}{|C_{\sigma_1}|} = \sum_{\sigma_1 \in S_d} \frac{1}{|C_{\sigma_1}|} = c, \qquad (7.13)$$

where c is the number of conjugacy classes in S_d.

Exercise 7.4.4. Let $\lambda = (\lambda_1^{k_1}, \lambda_2^{k_2}, \ldots, \lambda_n^{k_n})$ be a partition of d. The exponents in the notation mean that the integer λ_1 is repeated k_1 times in the partition, etc...for example, $\lambda = (3, 3, 3, 3, 2, 1, 1, 1) = (3^4, 2, 1^3)$. Prove that if σ is a permutation of cycle type λ, then

$$|\xi(\sigma)| = \prod_{i=1}^{n} \lambda_i^{k_i}(k_i!).$$

Example 7.4.5. We restrict our attention to degree 2 and compute connected and disconnected Hurwitz numbers for unramified covers of a genus g curve. In this case the Riemann–Hurwitz formula determines the genus of the cover curve to be $2g - 1$.

We begin with the disconnected case. The fundamental group of a genus g surface is presented with $2g$ generators $\{\alpha_i, \beta_i\}_{i=1,\ldots,g}$ subject to the relation that the product of commutators $\prod_{i=1}^{g}[\alpha_i, \beta_i]$ is trivial.

A monodromy representation of type $(g, 2, \varnothing)$ consists of an arbitrary choice of $2g$ elements of S_2: since S_2 is abelian, the relation on the generators of the fundamental group imposes no restriction on the choices of images of the generators. This yields 2^{2g} monodromy representations, and

$$H^{\bullet}_{2g-1 \xrightarrow{2} g}(\varnothing) = 2^{2g-1}. \qquad (7.14)$$

There is only one monodromy representation that gives rise to a disconnected cover: when all generators are sent to the identity element we obtain a trivial double cover of the base curve. Hence we have

$$H_{2g-1 \xrightarrow{2} g}(\varnothing) = \frac{2^{2g} - 1}{2}. \qquad (7.15)$$

We leave it as an exercise to introduce ramification conditions for degree 2 covers and compute the general Hurwitz number of degree 2.

Exercise 7.4.5. Compute all Hurwitz numbers of degree 2.

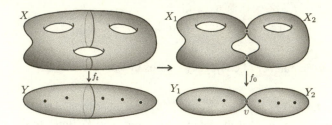

Figure 7.7 Degeneration of a cover to a nodal cover. Source and target degenerate simultaneously and the ramification orders on both sides of the node match.

7.5 Degeneration Formulas

Hurwitz numbers exhibit interesting recursive structure, i.e. a general Hurwitz number may be expressed (typically via combinatorially involved formulas) in terms of Hurwitz numbers where some of the discrete data (genus, degree or number of partitions) is smaller.

To get an intuition as to why that may be the case, look at Figure 7.7; on the left-hand side we have drawn a Hurwitz cover and identified a loop on the base curve and its inverse image on the cover. We shrink simultaneously all such loops to obtain a cover of a nodal curve by a nodal curve. We have not discussed singular curves much in this book, but, as the picture suggests, a cover of the curve $Y_1 \cup Y_2$ consists of a pair of covers $(X_1 \to Y_1, X_2 \to Y_2)$ subject to the condition that the ramification profile over the points that get glued together must be equal. Observe that the genera of X_1 and X_2 are strictly less than the genus of X, and that, even with the addition of the point that gets glued, the number of branch points on Y_1 and Y_2 is less than it was on Y. It then seems that one should be able to express the total number of Hurwitz covers of X as a sum of products of Hurwitz covers of Y_1 and Y_2: this is exactly what the degeneration formulas do.

Theorem 7.5.1 (Degeneration Formulas). *For $v \vdash d$ a partition of d, denote $|\xi(v)|$ the order of the centralizer of any permutation in S_d of cycle type v. Then the following formulas hold for all Hurwitz data.*

1. Base curve of genus 0: reducing branch points

$$H^{\bullet}_{g \overset{d}{\to} 0}(\lambda_1, \ldots, \lambda_s, \mu_1, \ldots, \mu_t)$$
$$= \sum_{v \vdash d} |\xi(v)| H^{\bullet}_{g_1 \overset{d}{\to} 0}(\lambda_1, \ldots, \lambda_s, v) H^{\bullet}_{g_2 \overset{d}{\to} 0}(v, \mu_1, \ldots, \mu_t);$$

g_1 and g_2 are determined by the Riemann–Hurwitz formula and they satisfy the condition $g_1 + g_2 + \ell(v) - 1 = g$.

2. Reducing the genus of a higher genus base curve

$$H^{\bullet}_{\underset{h \xrightarrow{d} g}{}} (\lambda_1, \ldots, \lambda_s) = \sum_{v \vdash d} |\xi(v)| H^{\bullet}_{\underset{h - \ell(v) \xrightarrow{d} g - 1}{}} (v, v, \lambda_1, \ldots, \lambda_s).$$

Remarks 7.5.2.

1. One may obtain many different degeneration formulas depending on the genus of the base curve and on the choice of the loop to shrink. We choose to write and highlight only two types, to keep the exposition cleaner. These will be sufficient for the reconstruction theorem (Theorem 7.5.3) we intend to prove.

2. A direct geometric proof of the degeneration formulas is quite complicated, as several Hurwitz covers degenerate to the same nodal cover. However, keeping track of this phenomenon becomes relatively simple when we describe covers via their monodromy representations.

3. There are degeneration formulas expressing connected Hurwitz numbers in terms of connected Hurwitz numbers, but they are more complicated: note that when you degenerate a connected cover X to a nodal cover $X_1 \cup X_2$, X_1 or X_2 may be disconnected even if their union is connected. Then to express such a contribution in terms of connected Hurwitz numbers, one has to keep track of the Hurwitz data on each individual connected component of the degenerated cover.

Proof By Theorem 7.3.2 we have

$$d! H^{\bullet}_{\underset{g \xrightarrow{d} 0}{}} (\lambda_1, \ldots, \lambda_s, \mu_1, \ldots, \mu_t) = |M|, \qquad (7.16)$$

where M is the set of monodromy representations of appropriate type. An element in M consists of a tuple of permutations $(\sigma_1, \ldots, \sigma_s, \tilde{\sigma}_1, \ldots, \tilde{\sigma}_t)$, such that the permutation σ_i has cycle type λ_i, the permutation $\tilde{\sigma}_j$ has cycle type μ_j, and the product of all permutations is the identity. Consider the set:

$$Q := N_{\lambda, v} \times N_{v, \mu}$$

where an element of $N_{\lambda, v}$ is a tuple $(\sigma_1, \ldots, \sigma_s, \pi_1)$ where σ_i has cycle type λ_i, π_1 has cycle type v, and the product of all permutations is the identity. Similarly an element of $N_{v, \mu}$ is a tuple $(\pi_2, \tilde{\sigma}_1, \ldots, \tilde{\sigma}_t)$, with π_2 of cycle type v, $\tilde{\sigma}_j$ has cycle type μ_j and the product of all permutation equals the identity. Let $P \subseteq Q$ denote the subset of Q where $\pi_1 = \pi_2^{-1}$.

Exercise 7.5.1. The cardinality of the set P is

$$|P| = \sum_{v \vdash d} \frac{1}{|C_v|} |N_{\lambda,v}||N_{v,\mu}|, \tag{7.17}$$

where C_v denotes the conjugacy class of permutations of cycle type v.

Given an element $(\sigma_1, \ldots, \sigma_s, \tilde{\sigma}_1, \ldots, \tilde{\sigma}_t) \in M$, we can produce the element $(\sigma_1, \ldots, \sigma_s, (\prod \sigma_i)^{-1}), (\prod \sigma_i, \tilde{\sigma}_1, \ldots, \tilde{\sigma}_t) \in P$. Conversely, to the pair of tuples $(\sigma_1, \ldots, \sigma_s, \pi_1), (\pi_1^{-1}, \tilde{\sigma}_1, \ldots, \tilde{\sigma}_t) \in P$ we can assign $(\sigma_1, \ldots, \sigma_s, \tilde{\sigma}_1, \ldots, \tilde{\sigma}_t) \in M$. The two functions just constructed are inverses of each other, which shows that the cardinalities of P and M are equal. Combining this fact with (7.16) and (7.17), we obtain

$$d! H^{\bullet}_{g \xrightarrow{d} 0}(\lambda_1, \ldots, \lambda_s, \mu_1, \ldots, \mu_t)$$

$$= \sum_{v \vdash d} \frac{1}{|C_v|} d! H^{\bullet}_{g_1 \xrightarrow{d} 0}(\lambda_1, \ldots, \lambda_s, v) d! H^{\bullet}_{g_2 \xrightarrow{d} 0}(v, \mu_1, \ldots, \mu_t).$$

The first degeneration formula follows by using the identity $|C_v||\xi(v)| = d!$.

The second formula is proven similarly. Denote M the set of monodromy representations computing $H^{\bullet}_{h \xrightarrow{d} g}(\lambda_1, \ldots, \lambda_s)$. Denote by N_v the set of monodromy representations for $H^{\bullet}_{h-\ell(v) \xrightarrow{d} g-1}(v, v, \lambda_1, \ldots, \lambda_s)$. The set M and all the sets N_v are identified with sets of tuples of permutations satisfying the relation coming from the standard presentation of the fundamental group of a punctured surface seen in (5.8) (only remember to write the relation from right to left because now we are multiplying permutations).

There is a natural function

$$M \to \bigcup_{v \vdash d} N_v,$$

defined by

$$(\alpha_1, \beta_1, \ldots, \alpha_g, \beta_g, \sigma_1, \ldots, \sigma_s)$$

$$\mapsto (\alpha_1, \beta_1, \ldots, \alpha_{g-1}, \beta_{g-1}, (\alpha_g^{-1}\beta_g\alpha_g), \beta_g^{-1}, \sigma_1, \ldots, \sigma_s). \tag{7.18}$$

Let us turn our attention to what is actually happening here. Observe that both the fundamental group of a genus g surface with n punctures and the fundamental group of a genus $g - 1$ surface with $n + 2$ punctures have $2g + n$ generators. There are generators corresponding to the sides of an identification polygon for the surface, which we call *genus generators*, and *puncture generators* winding around each of the punctures.

In the fundamental groups on either side of (7.18) we are exchanging two genus generators with two puncture generators. If the image of the two genus generators to be eliminated are α_g and β_g, we assign to the new puncture generators images $(\alpha_g^{-1}\beta_g\alpha_g)$ and β_g^{-1}. This way before and after, the relation imposed from the presentation of the fundamental groups is the same:

$$\sigma_s \ldots \sigma_1 \beta_g^{-1}\alpha_g^{-1}\beta_g\alpha_g \ldots \beta_1^{-1}\alpha_1^{-1}\beta_1\alpha_1 = e.$$

We finally note that the images of the two new puncture generators are conjugate to each other; hence, for any tuple in M its image must be in one of the N_ν.

The function (7.18) is surjective: given a tuple $(\ldots, \beta_{g-1}, \pi_1, \pi_2, \sigma_1 \ldots) \in N_\nu$, its inverse image consists of tuples where $\beta_g = \pi_2^{-1}$, and α_g satisfies the equation

$$\alpha_g^{-1}\pi_2^{-1}\alpha_g = \pi_1.$$

Since π_1 and π_2 are in the same conjugacy class, the equation has precisely $|\xi(\nu)|$ solutions.

Now we can count the cardinality of M by adding the cardinality of all inverse images via (7.18), to obtain

$$|M| = \sum_{\nu \vdash d} |\xi(\nu)||N_\nu|.$$

The second degeneration formula follows by replacing $|M|$ and the $|N_\nu|$ with the corresponding Hurwitz numbers (times $d!$, that cancels). □

The degeneration formulas imply the following interesting reconstruction theorem.

Theorem 7.5.3. *All (disconnected) degree d Hurwitz numbers are determined in terms of Hurwitz numbers of the form $H^\bullet_{\substack{d \\ h \to 0}}(\lambda_1, \lambda_2, \lambda_3)$, i.e. where the genus of the base curve is 0 and there are only three ramification conditions imposed.*

Proof Start with a general Hurwitz number $H^\bullet_{\substack{d \\ h \to g}}(\lambda_1, \ldots, \lambda_s)$. By applying g times the second degeneration formula, this number will be written as a (monstrous) sum of Hurwitz numbers where the genus of the base curve is 0 and now there are $s + 2g$ ramification conditions.

Now consider a loop around two of the branch points and apply the first degeneration formula: the Hurwitz number will then be written in terms of products of one Hurwitz number with exactly three branch points and another with $s + 2g - 1$ branch points. Iterating this procedure $s + 2g - 4$ more times, we obtain an expression for our original Hurwitz number in terms of a (monstrous) sum of (monstrous) products of genus 0, three-pointed Hurwitz numbers.   □

Exercise 7.5.2. Compute the Hurwitz number $H_{2 \overset{2}{\to} 1}((2), (2)) = 2$ using the degeneration formulas.

8
Representation Theory of S_d

One could argue that representation theory is a branch of mathematics devoted to translating group theory into linear algebra. Informally, a representation of an abstract group G is a homomorphism from G to a group of matrices. The name comes from the fact that the above group of matrices is a concrete representative for the isomorphism class of G. Then matrices correspond to linear transformations of vector spaces, and therefore G may be viewed as a collection of transformations of Euclidean space. If you have ever thought of a cyclic group of order n as the group of rotations in the plane that preserve a regular n-gon centered at the origin, you have actually thought about a representation of the cyclic group! Historically, this is actually how groups were born: in Felix Klein's 1884 *Lectures on the Icosahedron* (Klein, 1956), one may see many concepts of modern group theory arising via the study of groups of symmetries of shapes, i.e. groups of linear transformations of two- or three-dimensional space which preserve a given shape.

Since then, representation theory has evolved into a vast, far-reaching and quite sophisticated area of mathematics. Here we wish to give an essential introduction to some of the ideas that are important in our next translation of the Hurwitz problem. Our goal is to come as efficiently as possible to understand the class algebra of the symmetric group S_d. For this reason we choose to make most of our exposition specific to the symmetric group S_d, and state without proof many facts where we feel the proof would not be especially relevant to our Hurwitz story. The reader interested in finding proofs and filling in more details may consult Dummit and Foote (2004, Chapter 18) or Fulton and Harris (1991, Part 1).

8.1 The Group Ring and the Group Algebra

A natural step in trying to convert group theory into linear algebra information is to construct a vector space that "knows a lot" about the group. We construct

111

a ring that encodes the group operation as its multiplication, and then enlarge coefficients to also have a vector space structure.

Definition 8.1.1. The **group ring** of the symmetric group S_d, denoted $\mathbb{Z}[S_d]$, as a set, consists of all formal $\mathbb{Z}$-linear combinations of elements of S_d:

$$\mathbb{Z}[S_d] = \left\{ \sum_{\sigma \in S_d} a_\sigma \sigma \mid a_\sigma \in \mathbb{C} \right\}.$$

Addition is formal:

$$\sum_{\sigma \in S_d} a_\sigma \sigma + \sum_{\sigma \in S_d} b_\sigma \sigma = \sum_{\sigma \in S_d} (a_\sigma + b_\sigma)\sigma.$$

Multiplication for elements of the form σ is defined to be the multiplication of S_d, and is extended to arbitrary linear combination by requiring it to be bilinear.

Since S_d is not an abelian group, $\mathbb{Z}[S_d]$ is not a commutative ring.

Example 8.1.2. For $G = S_3$, $x = 3(12) + 5(123)$ and $y = 4(13) - 6(123) = 4(13) + (-6)(123)$ are elements of $\mathbb{Z}[S_3]$. We have

$$x + y = 3(12) + 4(13) + (5 - 6)(123) = 3(12) + 4(13) - (123)$$

and

$$
\begin{aligned}
x \cdot y &= (3(12) + 5(123))(4(13) - 6(123)) & (8.1)\\
&= 12(12)(13) - 18(12)(123) + 20(123)(13) - 30(123)(123) & (8.2)\\
&= 12(132) - 18(23) + 20(23) - 30(132) & (8.3)\\
&= 2(23) - 18(132). & (8.4)
\end{aligned}
$$

Example 8.1.2 illustrates that carrying out the product of elements of $\mathbb{Z}[S_d]$ happens in two steps: first one uses bilinearity to go from a product of sums to a sum of products of permutations (8.2); then one uses the group operation to reduce each of the products to just one permutation (8.3).

We call the expression in (8.2) the **formal expansion** of a product of elements of $\mathbb{Z}[S_d]$, and call each of the terms in the sum an **ordered monomial** in the formal expansion. Order matters since multiplication is not commutative.

Definition 8.1.3. We denote by $\mathbb{C}[S_d]$ the set of formal linear combinations of group elements where the coefficients a_σ are complex numbers. Together with addition and multiplication defined as above, there is a natural way to multiply elements of $\mathbb{C}[S_d]$ by scalars $t \in \mathbb{C}$:

$$t\left(\sum_{\sigma \in S_d} a_\sigma \sigma\right) = \sum_{\sigma \in S_d} (ta_\sigma)\sigma,$$

which also gives $\mathbb{C}[S_d]$ the structure of a vector space, with a natural basis given by $\sigma \in S_d$. A set with operations that make it simultaneously a ring and a vector space is called an **algebra**, and $\mathbb{C}[S_d]$ is called the **group algebra** of S_d.

8.2 Representations

We define representations in three equivalent ways – and encourage the reader to become familiar with all points of view, as each has its own advantages.

Definition 8.2.1. A (finite-dimensional, complex) **representation** ρ of S_d is, equivalently:

(group action) a finite-dimensional vector space V_ρ together with a linear action of S_d, i.e. a map

$$\text{"dot"} : S_d \times V_\rho \to V_\rho$$

such that, for every $\sigma, \sigma_1, \sigma_2 \in S_d$, $v, w \in V_\rho$, $\lambda \in \mathbb{C}$:

1 $e \cdot v = v$;
2 $\sigma_2 \cdot (\sigma_1 \cdot v) = (\sigma_2 \sigma_1) \cdot v$;
3 $\sigma \cdot (v + w) = \sigma \cdot v + \sigma \cdot w$;
4 $\sigma \cdot \lambda v = \lambda \sigma \cdot v$.

(module) A finitely generated module over the group ring $\mathbb{C}[S_d]$, i.e. a finitely generated abelian group M_ρ together with a scalar multiplication $\star : \mathbb{C}[S_d] \times M_\rho \to M_\rho$ which distributes with respect to the algebra and group operations (see Dummit and Foote (2004, Chapter 10) for an explicit spelling out of the module axioms).

(homomorphism) A group homomorphism

$$\Phi_\rho : S_d \to GL(n, \mathbb{C}).$$

Exercise 8.2.1. Show that the three definitions are in fact equivalent.

Exercise 8.2.2. In mathematics, we follow the philosophy that for a class of mathematical objects with certain structure and properties (e.g. operations satisfying axioms), the "desirable" functions are those which respect all of the structure and properties used to define the objects. Functions that pass this test are then called **morphisms**, and invertible morphisms are called **isomorphisms**. Develop the appropriate notions of morphism and isomorphism for representations.

The dimension of V_ρ (or the rank of M_ρ, or the n in $GL(n, \mathbb{C})$) is called the **dimension** of the representation ρ.

A **subrepresentation** $\rho' \leq \rho$ is an invariant subspace (or a $\mathbb{C}[S_d]$ submodule) $U_{\rho'}$ of V_ρ. The 0 vector and V_ρ itself are trivial examples of subrepresentations of ρ. A representation ρ that does not contain any nontrivial subrepresentation is called **irreducible**.

Example 8.2.2. The one-dimensional vector space $\mathbb{C}$ can be made into a representation by letting S_d act trivially:

$$\sigma \cdot z = z,$$

for all $\sigma \in S_d$, $v \in V$. This is called the **trivial** representation of S_d and denoted ρ_1. The trivial representation is irreducible simply because a one-dimensional vector space does not have any proper subspace except $\{0\}$.

Example 8.2.3. The group action on $\mathbb{C}$:

$$\sigma \cdot z = \begin{cases} z & \text{if } \sigma \text{ is an even permutation} \\ -z & \text{if } \sigma \text{ is an odd permutation} \end{cases}$$

gives another one-dimensional (hence irreducible) representation of S_d, called the **sign** representation, denoted ρ_{-1}.

Example 8.2.4. Consider a d-dimensional vector space V with basis $\{e_1, \ldots, e_d\}$. Define a group action of S_d on V by extending by linearity the following action on the bases vectors:

$$\sigma \cdot e_i = e_{\sigma(i)}.$$

This is called a **permutation** representation. We see that it is not an irreducible representation by noting that the linear span of the vector $e_1 + \ldots + e_d$ is invariant under the action of S_d (and hence it gives a proper subrepresentation isomorphic to the trivial representation).

Exercise 8.2.3. Refer to the module axioms from Dummit and Foote (2004, Section 10.1) and prove that the group ring itself is in a natural way a module over itself. As such, it is a representation of S_d called the **regular representation**.

An important feature of the regular representation is that it contains all irreducible representations of S_d: we make this statement precise in the next paragraph.

Exercise 8.2.4. Given two representations ρ_1 and ρ_2, one can form the **direct sum** representation $\rho_1 \oplus \rho_2$ by considering the direct sum of the corresponding vector spaces $V_{\rho_1} \oplus V_{\rho_2}$ with the natural extension of the action. Describe the matrix $\Phi_{\rho_1 \oplus \rho_2}(\sigma)$ in terms of $\Phi_{\rho_1}(\sigma)$ and $\Phi_{\rho_2}(\sigma)$.

We recall a few fundamental facts about representations (Fulton and Harris, 1991):

1. Any finite-dimensional representation of S_d decomposes uniquely (up to the order of the factors) as a direct sum of irreducible representations;
2. The number of irreducible representations of S_d equals the number of conjugacy classes of S_d, which in turn are naturally indexed by partitions of the integer d;
3. Denote by ρ an irreducible representation of S_d, and V_ρ the corresponding vector space, and understand a sum over the index ρ to mean to sum over all irreducible representations of S_d. Then the regular representation decomposes as

$$\mathbb{C}[S_d] \cong \bigoplus_\rho V_\rho^{\oplus \dim \rho}. \tag{8.5}$$

By equating the dimensions on either side of (8.5), we obtain

$$d! = \sum_\rho (\dim \rho)^2. \tag{8.6}$$

Example 8.2.5. We describe all irreducible representations of S_3. By point 2 above, there are three irreducible representations. We already know two of them: the trivial and the sign representations. It then follows from (8.6) that the last irreducible representation must be two-dimensional. This is called the **standard** representation and denoted by ρ_S. One way to construct the standard representation is to consider the quotient vector space of the three-dimensional permutation representation of S_3 by the invariant line $\langle e_1 + e_2 + e_3 \rangle$, and notice that the permutation action naturally descends, since we are quotienting by a trivial representation.

Exercise 8.2.5. Convince yourself that the standard representation is irreducible by showing that it does not admit either the trivial or the sign representation as a subrepresentation. In Section 8.3 we see a much more efficient way to prove the irreducibility of ρ_S.

8.3 Characters

Characters should be thought as "coordinates" for representations. They allow us to describe representations via a finite list of numbers which *play well* with many algebraic constructions in representation theory. We devote this section to making this statement more precise and recalling some basic notions and properties of characters.

Definition 8.3.1. Let ρ be a representation of S_d. The **character** of ρ is the function

$$\chi_\rho : S_d \to \mathbb{C}$$

defined as

$$\chi_\rho(\sigma) := \mathrm{trace}(\Phi_\rho(\sigma)).$$

The trace of a matrix is a coefficient of the characteristic polynomial of the associated linear transformation, and therefore it is invariant under conjugation (see Axler (1997, Chapters 9 and 10) for details).

This fact has two very important consequences:

1. The character of a representation does not depend on the choice of a basis for V_ρ (which gives rise to the matrices $\Phi_\rho(\sigma)$);
2. Characters are constant along conjugacy classes; functions with this property are called **class functions**.

A key fact about characters is that two complex representations of S_d are isomorphic if and only if they have the same character function[1] (Fulton and Harris, 1991). Therefore we can use characters to identify representations.

Exercise 8.3.1. Prove the following useful properties of characters:

1. For any representation ρ,

$$\chi_\rho(e) = \dim \rho. \tag{8.7}$$

---

[1] It is always true that isomorphic representations have the same characters. The converse statement requires us to be working over a field of characteristic 0.

Table 8.1 *The character tables of S_3 and S_4*

S_4	C_e	$C_{(2,1,1)}$	$C_{(2,2)}$	$C_{(3,1)}$	$C_{(4)}$
ρ_1	1	1	1	1	1
ρ_{-1}	1	-1	1	1	-1
ρ_2	2	0	2	-1	0
ρ_{3a}	3	-1	-1	0	1
ρ_{3b}	3	1	-1	0	-1

S_3	C_e	$C_{(2,1)}$	$C_{(3)}$
ρ_1	1	1	1
ρ_{-1}	1	-1	1
ρ_S	2	0	1

2. For any ρ_1, ρ_2,

$$\chi_{\rho_1 \oplus \rho_2} = \chi_{\rho_1} + \chi_{\rho_2}. \tag{8.8}$$

Exercise 8.3.2. Compute the character of the standard representation of S_3 as follows. First compute the character of the permutation representation ρ_P by explicitly constructing the relevant matrices. Then use (8.8) together with the fact that the permutation representation decomposes as $\rho_P \cong \rho_1 \oplus \rho_S$. In addition, use the characters $\chi_{\rho_1}, \chi_{\rho_{-1}}, \chi_{\rho_S}$ and (8.8) to show that ρ_S is irreducible.

Remark 8.3.2. Because the matrices $\Phi_\rho(\sigma)$ have finite order, it follows that characters take value in algebraic integers (sums of complex roots of unity). In the case of the symmetric group, characters are actually integer-valued functions.

We recall one final fact in our whirlwind tour of character theory. There is a complex inner product on the vector space of class functions, defined as follows:

$$\langle \alpha, \beta \rangle = \frac{1}{d!} \sum_{\sigma \in S_d} \alpha(\sigma) \overline{\beta(\sigma)}. \tag{8.9}$$

Characters of irreducible representations form an orthonormal basis for the vector space of class functions:

$$\langle \chi_{\rho_1}, \chi_{\rho_2} \rangle = \begin{cases} 1 & \rho_1 \cong \rho_2 \\ 0 & \rho_1 \ncong \rho_2 \end{cases} \tag{8.10}$$

for ρ_1 and ρ_2 irreducible.

Exercise 8.3.3. Check the orthonormality of the characters of the irreducible representations of S_3 and S_4. The characters are collected in Table 8.1.

8.4 The Class Algebra

We now introduce a commutative subalgebra of $\mathbb{C}[S_d]$, which plays a prominent role in our story.

Definition 8.4.1. The **class algebra** of S_d is the center of the group ring,

$$\mathcal{Z}\mathbb{C}[S_d] = \{x \in \mathbb{C}[S_d] | yx = xy \text{ for all } y \in \mathbb{C}[S_d]\}.$$

Exercise 8.4.1. For $\lambda \vdash d$ (a partition of the positive integer d), denote by $C_\lambda \in \mathbb{C}[S_d]$ the sum of all elements of cycle type λ.

1. Show that C_λ consists of the sum of all permutations in a particular conjugacy class;
2. Prove that for any λ, $C_\lambda \in \mathcal{Z}\mathbb{C}[S_d]$;
3. Show that the C_λs form a basis for $\mathcal{Z}\mathbb{C}[S_d]$ as a vector space:

$$\mathcal{Z}\mathbb{C}[S_d] = \bigoplus_{\lambda \vdash d} \langle C_\lambda \rangle_\mathbb{C}.$$

Hint: For $x \in \mathcal{Z}\mathbb{C}[S_d]$ we have $\sigma x \sigma^{-1} = x$ for any $\sigma \in S_d \subset \mathbb{C}[S_d]$. Now consider the sum

$$\sum_{\sigma \in S_d} \sigma x \sigma^{-1}.$$

We denote the conjugacy class of the identity element and the corresponding element in the class algebra by $C_e = C_{(1,\dots,1)} = e$.

The conjugacy class basis is a natural basis for $\mathcal{Z}\mathbb{C}[S_d]$. However, there is another basis, naturally indexed by the irreducible representations of S_d, that has a very nice multiplicative structure.

Theorem 8.4.2 (Maschke). *The class algebra $\mathcal{Z}\mathbb{C}[S_d]$ is a semisimple algebra, i.e. there is a basis $\{e_{\rho_1}, \dots, e_{\rho_n}\}$ (where the ρ_is are all irreducible representations of S_d) of idempotent elements. This means:*

$$e_{\rho_i} \cdot e_{\rho_j} = \begin{cases} e_{\rho_i} & \text{if } e_{\rho_i} = e_{\rho_j} \\ 0 & \text{otherwise.} \end{cases} \qquad (8.11)$$

Furthermore, the following changes of basis formulas hold

$$e_\rho = \frac{\dim \rho}{d!} \sum_\lambda \chi_\rho(\lambda) C_\lambda \qquad C_\lambda = |C_\lambda| \sum_\rho \frac{\chi_\rho(\lambda)}{\dim \rho} e_\rho \qquad (8.12)$$

where the summation index λ denotes all partitions λ of d, and the summation index ρ denotes all irreducible representations of S_d.

Example 8.4.3. The class algebra $\mathcal{Z}\mathbb{C}[S_3]$ is a three-dimensional vector space, with basis

$$C_e = e$$
$$C_{(2,1)} = (12) + (13) + (23)$$
$$C_{(3)} = (123) + (132).$$

The multiplication table of $\mathcal{Z}\mathbb{C}[S_3]$ is (generated bilinearly from)

	C_e	$C_{(2,1)}$	$C_{(3)}$
C_e	C_e	$C_{(2,1)}$	$C_{(3)}$
$C_{(2,1)}$	$C_{(2,1)}$	$3(C_e + C_{(3)})$	$2C_{(2,1)}$
$C_{(3)}$	$C_{(3)}$	$2C_{(2,1)}$	$2C_e + C_{(3)}.$

We denote the vectors of the semisimple basis for $\mathcal{Z}\mathbb{C}[S_d]$ by e_1, e_{-1} and e_S (instead of e_{ρ_1}, etc.). The changes of basis from Theorem 8.4.2 are:

$$e_1 = \tfrac{1}{6}(C_e + C_{(2,1)} + C_{(3)}) \qquad C_e = e_1 + e_{-1} + e_S$$

$$e_{-1} = \tfrac{1}{6}(C_e - C_{(2,1)} + C_{(3)}) \qquad C_{(2,1)} = 3e_1 - 3e_{-1} \qquad (8.13)$$

$$e_S = \tfrac{1}{3}(2C_e - C_{(3)}) \qquad C_{(3)} = 2e_1 + 2e_{-1} - e_S.$$

Exercise 8.4.2. Compute the change of basis between the semisimple and the conjugacy classes bases in the class algebra of the symmetric group S_4. Refer to Table 8.1 for the characters of S_4.

9

Hurwitz Numbers and $\mathcal{Z}(S_d)$

In this chapter we make yet another translation of the Hurwitz problem: when the fundamental group of a punctured surface is presented by a finite set of generators, then a monodromy representation is equivalent to a choice of elements in S_d: the images of each generator via the monodromy representation. Each such element is required to belong to a specified conjugacy class, and the totality of the elements must satisfy the relations coming from the presentation of the fundamental group. The number of monodromy representations, which becomes the number of all possible choices of such elements, can then be viewed as a particular coefficient of a product of elements in the class algebra $\mathcal{Z}\mathbb{C}[S_d]$.

We then exploit the fact that the class algebra of the symmetric group is semisimple (Theorem 8.4.2), i.e. it has a basis that makes multiplication idempotent: writing vectors in the semisimple basis makes a product of vectors into products of the coefficients of each individual basis vector. This allows us to write down closed formulas for arbitrary Hurwitz numbers, in terms of the coefficients of the change of basis, i.e. characters of the symmetric group. The 2006 Fields medalist Andrei Okounkov attributed these formulas to Burnside – as they appear as an exercise in his book *Theory of Groups of Finite Order* (Burnside, 1955). The Burnside formulas justify the slogan that Hurwitz theory is equivalent to character theory of S_d.

9.1 Genus 0

We begin our study by setting the genus of the base curve to be 0. In this case the fundamental group of a punctured sphere is a free group, and a monodromy representation is obtained by choosing elements in S_d that belong to specified conjugacy classes (Example 7.1.8). A concise way to express all such possible choices is given in the following proposition.

Proposition 9.1.1. *Let $\lambda_1, \ldots, \lambda_n$ be partitions of the integer d and for every i denote by $C_{\lambda_i} \in \mathcal{Z}\mathbb{C}[S_d]$ the basis element associated to the corresponding conjugacy class, i.e. the sum of all elements in S_d of cycle type λ_i. A disconnected, genus 0 Hurwitz number is given by*

$$H_{\substack{\bullet \\ h \to 0}}^{d}(\lambda_1, \ldots, \lambda_n) = \frac{1}{d!}[C_e]C_{\lambda_n} \ldots C_{\lambda_2}C_{\lambda_1},$$

where $[C_e]C_{\lambda_n} \ldots C_{\lambda_2}C_{\lambda_1}$ denotes the coefficient of $C_e = \{e\}$ after writing the product $C_{\lambda_n} \ldots C_{\lambda_2}C_{\lambda_1}$ as a linear combination of the basis elements $C_\lambda \in \mathcal{Z}\mathbb{C}[S_d]$. Note that the genus h of the cover curve is determined by the Riemann–Hurwitz Formula.

We give an example demonstrating this notation, then carry out the proof.

Example 9.1.2. In $\mathcal{Z}\mathbb{C}[S_3]$ we have

$$C_{(3)}C_{(3)} = ((123) + (132))((123) + (132)) \tag{9.1}$$

$$= (123)(132) + (132)(123) + (132)(132) + (123)(123) \tag{9.2}$$

$$= 2e + (123) + (132) = 2C_e + C_{(3)}. \tag{9.3}$$

Thus $[C_e]C_{(3)}C_{(3)} = 2$.

Proof of Proposition 9.1.1 We show that the number M of monodromy representations of type $(0, d, \lambda_1, \ldots, \lambda_n)$ is equal to $[C_e]C_{\lambda_n} \ldots C_{\lambda_2}C_{\lambda_1}$. Proposition 9.1.1 then follows from Theorem 7.3.1.

The fundamental group of $\mathbb{P}^1(\mathbb{C}) \smallsetminus \{b_1, \ldots, b_n\}$ is a free group in $n - 1$ generators, but it is conveniently presented in a more symmetric way as:

$$\pi_1 \left(\mathbb{P}^1(\mathbb{C}) \smallsetminus \{b_1, \ldots, b_n\} \right) \cong \langle \rho_1, \ldots, \rho_n | \rho_1 \rho_2 \cdots \rho_n \rangle,$$

where ρ_i is a symbol for the corresponding small loop around b_i. A monodromy representation Φ of type $(0, d, \lambda_1, \ldots, \lambda_n)$ is given by a choice of $\sigma_1, \ldots, \sigma_n \in S_d$ (the images via Φ of the small loops ρ_i) such that each σ_i has cycle type λ_i, and such that $\sigma_n \sigma_{n-1} \cdots \sigma_1 = e$.

Write down each C_{λ_i} as the formal sum of the group elements in the corresponding conjugacy class, and formally expand the product $C_{\lambda_n} \ldots C_{\lambda_2}C_{\lambda_1}$ (without actually multiplying any of the group elements – as in (9.2)). Each n-tuple $\sigma_1, \ldots, \sigma_n$ giving a monodromy representation appears uniquely as an ordered monomial in such expansion. Conversely, an ordered monomial in the expansion of $C_{\lambda_n} \ldots C_{\lambda_2}C_{\lambda_1}$ corresponds to a monodromy representation if the product of the corresponding group elements gives the identity.

There is therefore a natural bijection between monodromy representations of type $(0, d, \lambda_1, \ldots, \lambda_n)$ and terms in the formal expansion of $C_{\lambda_n} \ldots C_{\lambda_2} C_{\lambda_1}$ whose product is the identity. This precisely means that the number of monodromy representations equals the coefficient of C_e in the product $C_{\lambda_n} \ldots C_{\lambda_2} C_{\lambda_1}$. $\qquad\qquad\qquad\qquad\qquad\qquad\qquad\qquad\qquad\qquad\qquad$ □

Example 9.1.3. The Hurwitz number $H_{0 \to 0}^{\ 3}((3), (3)) = (1/3!)[C_e]C_{(3)} \cdot C_{(3)}$ with the product computed in $\mathcal{Z}\mathbb{C}[S_3]$. The coefficient $[C_e]C_{(3)} \cdot C_{(3)}$ is computed in Example 9.1.2 as 2, and hence $H_{0 \to 0}^{\ 3}((3), (3)) = 2/6 = 1/3$. (This is our third time computing this Hurwitz number! Take a walk down memory lane with Example 6.1.7 and Exercise 7.3.1.)

9.2 Genus and Commutators

We now give a formula for higher-genus Hurwitz numbers as a product of elements in the class algebra $\mathcal{Z}\mathbb{C}[S_d]$. The key idea here is that in the "standard" presentation of the fundamental group of a punctured positive genus surface, there are additional generators corresponding to loops winding around the handles, and the product of all generators now must satisfy a relation which contains products of commutators. Before we discuss how to translate all of this in the language of the class algebra, we observe a simple but illustrative example.

Example 9.2.1. The Hurwitz number $H_{1 \to 1}^{\bullet}{}_{d}(\varnothing)$ was computed in Example 7.4.4 as the number of irreducible representations, or of conjugacy classes, of S_d. Here we give an alternative formula:

$$H_{1 \to 1}^{\bullet}{}_{d}(\varnothing) = \frac{1}{d!}[C_e] \sum_{\lambda \vdash d} |\xi(\lambda)| C_\lambda^2, \qquad (9.4)$$

where $|\xi(\lambda)|$ is the size of the centralizer of any permutation in the conjugacy class indexed by λ.

To each monodromy representation contributing to $H_{1 \to 1}^{\bullet}{}_{d}$ we associate an ordered monomial in the formal expansion of (9.4) as follows. A monodromy representation Φ of type $(1, d, \varnothing)$ corresponds to a choice of $\sigma_1, \sigma_2 \in S_d$ such that $\sigma_2^{-1}\sigma_1^{-1}\sigma_2\sigma_1 = e$. We may think of all possible such choices as picking first an element $\sigma_1 \in S_d$, then picking a σ_2 that gives rise to a conjugate element $\hat{\sigma}_1 = \sigma_2^{-1}\sigma_1^{-1}\sigma_2$ such that $\hat{\sigma}_1\sigma_1 = e$. Any pair $\hat{\sigma}_1\sigma_1$ identifies an ordered monomial in the formal expansion of $\sum_{\lambda \vdash d} C_\lambda^2$ (see Example 8.1.2). The assignment $\Phi \mapsto \hat{\sigma}_1\sigma_1$ defines a function κ from the set of monodromy representations of type $(1, d, \varnothing)$ to the set of ordered monomials in the formal expansion of 9.4.

Exercise 9.2.1.

1. Show that the image of κ consists precisely of the ordered monomials whose product returns the identity element.
2. Prove that two permutations σ_2 and σ_2' give rise to the same $\hat{\sigma}_1$ if and only if $\sigma_2^{-1}\sigma_2' \in \xi(\sigma_1)$. Conclude that for any $\hat{\sigma}_1\sigma_1 \in C_\lambda^2$ in the image of κ, the cardinality of the inverse image $\kappa^{-1}(\hat{\sigma}_1\sigma_1)$ is $|\xi(\lambda)|$.

Formula (9.4) follows from computing the cardinality M of the set of monodromy representations as

$$M = \sum_{\hat{\sigma}_1\sigma_1 \in Im(\kappa)} |\kappa^{-1}(\hat{\sigma}_1\sigma_1)| = \sum_{\lambda \vdash d} |\xi(\lambda)| \left([C_e] C_\lambda^2 \right).$$

Example 9.2.1 is perhaps an overcomplicated way of computing a simple Hurwitz number, but it allowed us to introduce an important character in our story, which we now formally define.

Definition 9.2.2. For a fixed positive integer d we define

$$\mathfrak{K} := \sum_{\lambda \vdash d} |\xi(\lambda)| C_\lambda^2 \in \mathcal{Z}\mathbb{C}[S_d]. \tag{9.5}$$

The (Gothic) letter "k" is chosen from the German word *kommutator*: recall that the commutator of two elements $\sigma_1, \sigma_2 \in S_d$ is $[\sigma_1, \sigma_2] = \sigma_2^{-1}\sigma_1^{-1}\sigma_2\sigma_1$. One should think of $\mathfrak{K}$ as a way to express the sum of all commutators in S_d as an element in the class algebra $\mathcal{Z}\mathbb{C}[S_d]$. Exercise 9.2.2 provides the information needed to make this statement precise.

Exercise 9.2.2. Denote by X the set of ordered monomials in the formal expansion of the quadratic expression $\sum_{\lambda \vdash d} C_\lambda^2$. Consider the function[1]

$$\kappa : S_d \times S_d \to X$$

defined by $\kappa(\sigma_1, \sigma_2) = (\sigma_2^{-1}\sigma_1^{-1}\sigma_2)\sigma_1$. Show that κ is a surjective function and for every $\hat{\sigma}_1\sigma_1 \in X$, $|\kappa^{-1}(\hat{\sigma}_1\sigma_1)| = |\xi(\sigma_1)|$.

We are now ready to express a general (disconnected) Hurwitz number as a multiplication problem in the class algebra.

---

[1] Note that the function κ defined in Example 9.2.1 is the restriction of the κ defined here to the permutation pairs arising from monodromy representations.

Proposition 9.2.3. *Let* $\lambda_1, \ldots, \lambda_n$ *be partitions of the positive integer* d. *We have the formula*

$$H^{\bullet}_{\substack{d \\ h \to g}}(\lambda_1, \ldots, \lambda_n) = \frac{1}{d!}[C_e]\mathfrak{K}^g C_{\lambda_n} \ldots C_{\lambda_2} C_{\lambda_1}, \qquad (9.6)$$

where the genus h *of the cover curve is determined by the Riemann–Hurwitz formula.*

Proof   Recall from (5.7) that the fundamental group of a genus g surface with n punctures may be presented as:

$$\pi_1 \left(C_g \smallsetminus \{b_1, \ldots, b_n\} \right)$$
$$= \langle \rho_1, \ldots, \rho_n, \alpha_1, \beta_1, \ldots, \alpha_g, \beta_g | \rho_1 \ldots \rho_n [\alpha_1, \beta_1] \ldots [\alpha_g, \beta_g] \rangle,$$

where ρ_i is a small loop around b_i, α_j and β_j are the two independent loops around the j-th handle, and the square brackets denote commutators.

A monodromy representation of type $(g, d, \lambda_1, \ldots, \lambda_n)$ is then given by a choice of $\sigma_1, \ldots, \sigma_n, \mu_1, \ldots, \mu_g, \nu_1, \ldots, \nu_g \in S_d$ such that each σ_i has cycle type λ_i, and $[\mu_g, \nu_g] \cdots [\mu_1, \nu_1]\sigma_n \sigma_{n-1} \cdots \sigma_1 = e$.

Given a $(2g + n)$-tuple $\sigma_1, \ldots, \sigma_n, \mu_1, \nu_1, \ldots, \mu_g, \nu_g$ corresponding to a monodromy representation, one may write the ordered monomial

$$\hat{\mu}_g \mu_g \cdots \hat{\mu}_1 \mu_1 \sigma_n \sigma_{n-1} \cdots \sigma_1, \qquad (9.7)$$

where $\hat{\mu}_i = \nu_i^{-1} \mu_i^{-1} \nu_i$.

Monomials as in equation (9.7) appear in the formal expansion of the product $\left(\sum_{\lambda \vdash d} C_\lambda^2 \right)^g C_{\lambda_n} \ldots C_{\lambda_2} C_{\lambda_1}$ precisely as those terms where the product of the group elements gives the identity. Each such monomial arises from $\prod_{i=1}^{g} |\xi(\mu_i)|$ distinct monodromy representations. Recalling that $\mathfrak{K} = \sum_{\lambda \vdash d} |\xi(\lambda)| C_\lambda^2$, we see that the number of monodromy representations of type $(g, d, \lambda_1, \ldots, \lambda_n)$ corresponds to the coefficient of the identity in the product $\mathfrak{K}^g C_{\lambda_n} \ldots C_{\lambda_2} C_{\lambda_1}$, which concludes the proof of Proposition 9.2.3   □

9.3 Burnside Formula

Computing Hurwitz numbers is a multiplication problem in the class algebra of the symmetric group, and the conjugacy class basis $\{C_\lambda\}$ is well suited to encode the ramification profiles imposed over the branch points.

Theorem 8.4.2 shows that $\mathcal{Z}\mathbb{C}[S_d]$ is a semisimple algebra with a semisimple basis naturally indexed by irreducible representations. By changing basis we obtain a closed formula for Hurwitz numbers in terms of characters of the irreducible representations of S_d.

Theorem 9.3.1 (Burnside Character Formula). *Fix a positive integer d and m partitions $\lambda_i \vdash d$. Denote by ρ an irreducible representation of S_d, and understand a summation over the index ρ to be ranging over all irreducible representations. Then*

$$H^{\bullet}_{h \xrightarrow{d} g}(\lambda_1, \ldots, \lambda_m) = \sum_{\rho} \left(\frac{\dim \rho}{d!} \right)^{2-2g} \prod_{j=1}^{m} \frac{|C_{\lambda_j}| \chi_{\rho}(\lambda_j)}{\dim \rho}. \tag{9.8}$$

Remark 9.3.2. At first glance it might not be apparent why (9.8) represents an improvement over (9.6). Arguably, it is not: in mathematics, when we translate a problem we often just "shift" the complexity of the problem around. In formula (9.6) we have simple inputs (the conjugacy class basis vectors for $\mathcal{Z}\mathbb{C}[S_d]$), but we are multiplying vectors in a very high-dimensional algebra. In formula (9.8), the inputs are more sophisticated (the characters of representations of S_d), but the multiplication is now an ordinary multiplication of real numbers. In other words, we have shifted the complexity from the operation to the inputs.

Proof of Theorem 9.3.1 We first consider the element $\mathcal{R} \in \mathcal{Z}\mathbb{C}[S_d]$ and express it in terms of the semisimple basis. Using the change of basis formulas from Theorem 8.4.2, we have

$$\mathcal{R} = \sum_{\lambda} |\xi(\lambda)| C_{\lambda}^2 \tag{9.9}$$

$$= \sum_{\lambda} |\xi(\lambda)| \left(\sum_{\rho} \frac{|C_{\lambda}| \chi_{\rho}(\lambda)}{\dim \rho} e_{\rho} \right)^2 \tag{9.10}$$

$$= \sum_{\lambda} |\xi(\lambda)| \sum_{\rho} \left(\frac{|C_{\lambda}| \chi_{\rho}(\lambda)}{\dim \rho} \right)^2 e_{\rho} \tag{9.11}$$

$$= \sum_{\rho} \frac{d!}{(\dim \rho)^2} \left(\sum_{\lambda} |C_{\lambda}| \chi_{\rho}(\lambda)^2 \right) e_{\rho} \tag{9.12}$$

$$= \sum_{\rho} \frac{d!}{(\dim \rho)^2} \left(\sum_{\sigma \in S_d} \chi_{\rho}(\sigma)^2 \right) e_{\rho} \tag{9.13}$$

$$= \sum_{\rho} \left(\frac{d!}{\dim \rho} \right)^2 e_{\rho}. \tag{9.14}$$

In this string of equations we have applied the change of basis (9.10) and the orthonormality of the vectors e_ρ (9.11). Then we switched order of summation and used the identity $|C_\lambda||\xi(\lambda)| = d!$ to obtain (9.12), expressed the second summation as a sum over all elements of S_d (9.13), and finally recognized the second summation as the inner product of χ_ρ with itself (9.14).

Next we consider the product

$$
C_{\lambda_m} \cdots C_{\lambda_1} = \left(\sum_\rho \frac{|C_{\lambda_m}|\chi_\rho(\lambda_m)}{\dim \rho} e_\rho \right) \cdots \left(\sum_\rho \frac{|C_{\lambda_1}|\chi_\rho(\lambda_1)}{\dim \rho} e_\rho \right)
$$

$$
= \sum_\rho \prod_{j=1}^m \left(\frac{|C_{\lambda_j}|\chi_\rho(\lambda_j)}{\dim \rho} \right) e_\rho. \tag{9.15}
$$

Above, we see the magic of expressing a product of vectors in a semisimple basis: just multiply together the coefficients of each basis vector.

Incorporating (9.14) and (9.15) in formula (9.6)

$$
H^\bullet_{\underset{h \to g}{d}} (\lambda_1, \dots, \lambda_n) = \frac{1}{d!}[C_e]\aleph^g C_{\lambda_n} \cdots C_{\lambda_2} C_{\lambda_1}
$$

$$
= \frac{1}{d!}[C_e] \sum_\rho \left(\frac{d!}{\dim \rho} \right)^{2g} \prod_{j=1}^m \left(\frac{|C_{\lambda_j}|\chi_\rho(\lambda_j)}{\dim \rho} \right) e_\rho. \tag{9.16}
$$

We apply the inverse change of basis (back to the conjugacy class basis), and extract the coefficient of C_e. Recall from Theorem 8.4.2 that

$$
e_\rho = \frac{\dim \rho}{d!}\chi_\rho(e)C_e + \dots \tag{9.17}
$$

We observe that $\chi_\rho(e) = \dim \rho$; we finally obtain (9.8) by plugging (9.17) into (9.16). □

Example 9.3.3. Let us revisit the steps of the proof of Theorem 9.3.1 through the computation of $H_{\underset{1 \to 0}{3}}((3), (2, 1)^4)$. In this case the condition of a point with full ramification forces all covers to be connected, so $H = H^\bullet$. Refer to Table 8.1 for the character table of S_3 and the transformations from the conjugacy class basis to the representation basis. We have

$$
H_{\underset{1 \to 0}{3}}((3), (2, 1)^4) = \frac{1}{6}[C_e]C_{(3)}C^4_{(2,1)}
$$

$$
= \frac{1}{6}[C_e](2 \cdot 3^4 e_1 + 2 \cdot (-3)^4 e_{-1})
$$

$$
= \frac{1}{6} \left(\frac{2 \cdot 3^4}{6} + \frac{2 \cdot 3^4}{6} \right) = 9.
$$

Exercise 9.3.1. Compute the following Hurwitz numbers using the formula from Theorem 9.3.1:

1. $H_{2 \xrightarrow{3} 0}((3), (2, 1)^6)$;
2. $H_{5 \xrightarrow{3} 0}((3)^4, (2, 1)^6)$;
3. $H^{\bullet}_{0 \xrightarrow{3} 0}((2, 1)^4)$.

Now compute the general degree 3 disconnected Hurwitz number:

$$H^{\bullet}_{3g-2+a+b \xrightarrow{3} g}((3)^a (2, 1)^{2b}).$$

10

The Hurwitz Potential

We conclude our foray into Hurwitz theory by introducing some mathematical machinery which is useful to "organize things". Admittedly, there are lots of Hurwitz numbers... as in, infinitely many! But we have seen that they are not just some random collection of numbers unrelated to each other: we saw in Theorem 7.5.3 that they are all determined by base curve genus 0, three-branch-point Hurwitz numbers via recursive formulas.

It is sometimes convenient to consider infinite sets of numbers with some kind of recursive structure as coefficients of a power series, which is called a *generating function*. When the encoding is appropriate, operations on generating functions correspond to recursions on the collection of numbers.

We begin by introducing the notion of generating functions through some simple examples, which include the mind-boggling statement that there are "e" isomorphism classes of finite sets if, when we count them, we divide by the order of automorphism groups. Once we are warmed up, we introduce the *Hurwitz potential*, one ginormous power series that contains all Hurwitz numbers as coefficients of its monomials. We then derive two interesting applications of this point of view. The first is that the relationship between connected and disconnected Hurwitz numbers is controlled by one simple functional equation relating the connected and disconnected Hurwitz potentials. The second is that all (infinitely many!) recursions coming from a specific type of degeneration formula are encoded in a unique differential operator, called the *cut-and-join operator*, which vanishes when applied to the Hurwitz potential.

10.1 Generating Functions

The book (Wilf, 2006) introduces generating functions with this sentence:

> "A generating function is a clothesline on which we hang up a sequence of
> numbers for display."

Behind the humorous character of this statement lies the philosophy that encoding sequences of numbers as coefficients of power series is a convenient way to encode and manipulate combinatorial information. Let us make some precise definitions.

Definition 10.1.1. Given a sequence of numbers $A = \{a_n\}_{n \in \mathbb{Z}_{\geq 0}}$, the **ordinary generating function** for A is the formal power series:

$$\mathbf{f}(x) = \sum_{n \in \mathbb{Z}_{\geq 0}} a_n x^n. \tag{10.1}$$

The **exponential generating function** for A is defined to be

$$\mathbf{g}(x) = \sum_{n \in \mathbb{Z}_{\geq 0}} \frac{a_n}{n!} x^n. \tag{10.2}$$

If either of the above power series converges in a neighborhood of $x = 0$, then we also refer to the analytic function that the power series converges to as the generating function for A.

Example 10.1.2. If $A = \{1\}_{n \in \mathbb{Z}_{\geq 0}}$ is the constant sequence, then the ordinary and exponential generating functions for A are

$$\mathbf{f}(x) = \frac{1}{1-x} \qquad \mathbf{g}(x) = e^x. \tag{10.3}$$

Now, suppose that you don't know that A is the constant sequence, but you are rather told the recursive information that $a_0 = 1$ and for every nonnegative integer n, $a_n = a_{n+1}$. (I know, it must be a really bad day if you don't recognize a constant sequence from this information, but bear with us for the sake of exposition...) We may use this recursive information to prove the equations in (10.3) as follows.

We denote by $[x^n]\mathbf{f}$ the coefficient of x^n in the function $\mathbf{f}$ (which by Definition 10.1.1 is a_n, and we want to consider as an unknown) for every nonnegative integer n. The coefficient a_{n+1} can be viewed as the coefficient of x^n for the function $\frac{\mathbf{f}}{x}$. Therefore, for every $n \geq 0$, the recursion $a_n = a_{n+1}$ translates to the following equality of coefficients:

$$[x^n]\mathbf{f} = [x^n]\frac{\mathbf{f}}{x}. \tag{10.4}$$

We would like to remove the $[x^n]$ from (10.4) and thus replace infinitely many numerical equations with one functional equation. We must use caution, though, as the function $\frac{\mathbf{f}}{x}$ has a "pole" at 0, while $\mathbf{f}$ doesn't. We can fix this issue "by hand", i.e. by subtracting off the polar part from the right-hand side of (10.4) to obtain an honest functional equation satisfied by $\mathfrak{f}$:

$$\mathfrak{f} = \frac{\mathfrak{f}}{x} - \frac{1}{x}. \tag{10.5}$$

It is now immediate to solve (10.5) for $\mathfrak{f}$ and obtain (10.3).

For the exponential generating function, we have that for every $n \geq 0$,

$$[x^n]\frac{d\mathfrak{g}}{dx} = \frac{a_{n+1}}{n!} = \frac{a_n}{n!} = [x^n]\mathfrak{g},$$

and hence $\mathfrak{g}$ satisfies the ordinary differential equation $\mathfrak{g}' = \mathfrak{g}$ with boundary condition $\mathfrak{g}(0) = 1$. Recalling that first-order ODEs have unique solutions once the value of a point is specified, this uniquely determines $\mathfrak{g}(x) = e^x$.

Example 10.1.2 shows that a recursive relation among elements of a sequence can be turned into a functional or differential equation involving the corresponding generating functions. Whether the ordinary or exponential generating function encoding is more desirable often depends on the problem at hand. Sometimes in enumerative geometry the natural encoding is dictated to us from the geometric problem, as the next (also silly) example illustrates.

Example 10.1.3. We want to describe the generating function for the sequence counting *isomorphism classes of finite sets of cardinality n*. Isomorphisms of sets are just bijective functions, and any two sets of cardinality n can be put in bijection (that is essentially the definition of having cardinality n), so it looks like we are again just talking about the constant sequence. However we have already encountered in this book that enumerative geometers like to divide by the symmetries (i.e. the order of the automorphism group) of the geometric objects considered. A set of order n has S_n as its automorphism group: therefore the exponential generating function for the constant sequence is *morally* the best way to encode the counting of finite sets.

As a consequence, if you ask an enumerative geometer for the total number of finite sets, the answer will of course be:

$$\sum_{n=0}^{\infty} \frac{1}{n!} = e.$$

Sometimes we want to encode sets of numbers that depend on more than one index, and we do so by introducing a formal variable into the generating function for each index. We illustrate this philosophy with a simple example.

Example 10.1.4. Our counting problem is: $\{a_{k,n}\}$ = *number of subsets of order k of a set of cardinality n*. We introduce a formal variable q to keep track of the cardinality of the ambient set, and a variable x to keep track of the cardinality of the subset, and define

$$\mathsf{f}(x, q) = \sum_{n,k \geq 0} a_{n,k} x^k \frac{q^n}{n!}. \tag{10.6}$$

Note that we mixed things up a bit and chose to use exponential encoding for the variable q and ordinary encoding for x, for no particular reason other than to show that we can. We explore this generating function in the next two exercises.

Exercise 10.1.1. Show that each of the following equalities holds:

$$\mathsf{f}(x, q) = \sum_{n \geq 0} \frac{(x+1)^n q^n}{n!} = e^{(x+1)q}. \tag{10.7}$$

Hint: combine the fact from combinatorics that $a_{n,k} = \binom{n}{k}$ with Newton's binomial theorem.

Exercise 10.1.2. Show that setting $x = 1$ in (10.6) recovers the exponential generating function encoding the total number of subsets of a set of order n.

Example 10.1.4 and Exercise 10.1.2 illustrate a general philosophy: introducing new variables amounts to refining the combinatorial information, whereas specializing variables amounts to forgetting part of the structure.

Exercise 10.1.3. Let $p_0 = 1$ and p_n denote the number of partitions of the positive integer n; define the generating function for this sequence to be:

$$\mathsf{p}(x) := \sum_{n=0}^{\infty} p_n x^n = 1 + x + 2x^2 + 3x^3 + 5x^4 + 7x^5 + \ldots \tag{10.8}$$

Prove that

$$\mathsf{p}(x) = \prod_{k=1}^{\infty} \frac{1}{(1 - x^k)}. \tag{10.9}$$

We now introduce the generating function for Hurwitz numbers, which is what we are ultimately interested in. We begin by introducing a small shorthand notation for Hurwitz numbers.

Definition 10.1.5. For a positive integer r, we define

$$H^r_{h \xrightarrow{d} g} (\lambda_1, \ldots, \lambda_m) := H_{h \xrightarrow{d} g} (\lambda_1, \ldots, \lambda_m, (2, 1, \ldots, 1)^r).$$

In other words, we add r branch points corresponding to simple ramification to the ramification data.

The philosophy behind Definition 10.1.5 is that, besides the branch points with ramification profile specified by the partitions λ_i, we wish to allow some additional ramification, which we require to be generic.

Definition 10.1.6. The **genus g Hurwitz Potential** is a generating function for Hurwitz numbers counting covers of curves of genus g. We present it here with as many variables as possible. In almost all applications one makes a choice of the appropriate variables to maintain:

$$\mathcal{h}_g(p_{i,j}, u, z, q) := \sum H^r_{h \xrightarrow{d} g} (\lambda_1, \ldots, \lambda_m) \, p_{1,\lambda_1} \ldots p_{m,\lambda_m} \frac{u^r}{r!} z^{1-h} q^d,$$
(10.10)

where:

- For a partition $\lambda = (l_1, \ldots, l_k)$ the notation $p_{i,\lambda}$ is defined to mean:

$$p_{i,\lambda} = \prod_{j=1}^{l} p_{i,l_j},$$

where the variables $p_{i,j}$ index ramification profiles. The first index i keeps track of the branch point; the second gives one part of the partition denoting the ramification profile. Both i and j vary among all positive integers.
- u is a variable for the newly introduced additional simple ramification. The exponential encoding may seem a bit mysterious at this point, but we will soon see that it is very convenient.
- z indexes the genus of the cover curve (more precisely, it indexes $1/2$ the Euler characteristic, which is additive under disjoint unions).
- q keeps track of degree.

The **total Hurwitz potential** is defined to be the sum over all genera of the genus g potentials:

$$\mathcal{h} := \sum_{g=0}^{\infty} \mathcal{h}_g.$$
(10.11)

Similarly one can define disconnected Hurwitz potentials $\mathcal{h}_g^\bullet$, $\mathcal{h}^\bullet$, encoding disconnected Hurwitz numbers. Note that in the disconnected case the genus h of the cover curve ranges over all integers (see Definition 6.1.8).

Example 10.1.7. Hurwitz numbers do not depend on the ordering of the partitions specifying the ramification profiles, or equivalently on the labeling of the branch points.

However, in the Hurwitz potentials, the order is remembered. Thus a particular Hurwitz number appears as the coefficient of several different monomials. For example, the Hurwitz number $H^6_{3\to 0,3}((3)^2, (1,1,1))$ appears in $\mathcal{H}_0$ as the coefficient of the monomials:

$$p_{1,3}p_{2,3}p_{3,1}^3 \frac{u^6}{6!}z^{-4}q^3,$$

$$p_{1,3}p_{2,1}^3 p_{3,3}\frac{u^6}{6!}z^{-4}q^3,$$

and

$$p_{1,1}^3 p_{2,3}p_{3,3}\frac{u^6}{6!}z^{-4}q^3.$$

Example 10.1.8. Setting u and all the p variables equal to 0, we obtain a generating function $\mathcal{E}$ for étale (i.e. unramified) covers of Riemann Surfaces:

$$\mathcal{E}(z,q) = \mathcal{H}(p_{i,j}=0, u=0, z, q) = \sum_{d=1}^{\infty}\sum_{g=0}^{\infty} H^0_{dg-d+1\overset{d}{\to}g}(\varnothing)z^{d(1-g)}q^d$$

$$= q\frac{z^2}{z-1} + \frac{3}{2}q^2\frac{z^4}{(z^2-1)(z^2-4)} + \cdots \qquad (10.12)$$

Note that, because of the negative exponent used for the variable encoding the genus, the enumerative information is encoded in the coefficients of the Taylor/Laurent expansions of the above rational functions in z at $z = \infty$.

Exercise 10.1.4. Verify that the degree 1 and 2 terms of the étale Hurwitz potential $\mathcal{E}$ are given by the two summands in the last equation of (10.12). Refer to Example 7.4.5 for degree 2 étale Hurwitz numbers.

10.2 Connected Hurwitz Numbers

We first restrict our attention to connected unramified covers of genus 0 curves: we denote the generating function for such covers $\mathcal{E}_0$. The Riemann–Hurwitz formula tells us that the only possible unramified map to $\mathbb{P}^1(\mathbb{C})$ is a degree 1 map from $\mathbb{P}^1(\mathbb{C})$ itself. There is only one isomorphism class of such a map, and any map in the class has only the trivial automorphism. Hence

The Hurwitz Potential

$$\mathbb{C}_0(z, q) = zq. \tag{10.13}$$

Now allow the source curve to be possibly disconnected. For every degree d, there exists an unramified cover of $\mathbb{P}^1(\mathbb{C})$ of degree d: it consists of d distinct copies of $\mathbb{P}^1(\mathbb{C})$ (a genus $1 - d$ curve), each mapping down isomorphically to the base curve. There is one such map up to isomorphism and it has S_d as its automorphism group (since the d connected components of the source curve can be arbitrarily permuted). We therefore obtain

$$\mathbb{C}_0^\bullet(z, q) = \sum_{d=1}^{\infty} \frac{1}{d!} z^d q^d = e^{zq} - 1. \tag{10.14}$$

Thus, up to the constant term, the disconnected potential is the exponential of the connected potential. This is in fact a general phenomenon.

Theorem 10.2.1. *The connected and disconnected genus g Hurwitz potentials are related by exponentiation:*

$$1 + \mathcal{H}_g^\bullet = e^{\mathcal{H}_g}. \tag{10.15}$$

Proof The strategy of proof is to compare the coefficients of each individual monomial in (10.15). A monomial consists of fixing all discrete data for the Hurwitz problem. Once all the data are fixed, the disconnected Hurwitz number counts covers that can be organized as a sum over the discrete data for each of their connected components. For each of the summands, the contribution to the disconnected Hurwitz number consists of the product of the connected Hurwitz numbers divided by factorials corresponding to permuting connected components with identical discrete data. We now make this intuition precise. Unfortunately, the price to pay is having to set up some cumbersome notation.

We first observe that the q^0 coefficient (i.e. the constant term) of equation (10.15) is correct, since $[q^0]\mathcal{H}_g^\bullet = 0$.

Let $\mathfrak{d} = (h, d, r, \{\lambda_1, \ldots, \lambda_m\})$ denote the combinatorial data needed to identify a Hurwitz number for a base curve of genus g. We denote by $H_\mathfrak{d}$ (respectively $H_\mathfrak{d}^\bullet$) the Hurwitz number (respectively disconnected Hurwitz number) corresponding to such data. The Hurwitz number $H_\mathfrak{d}$ (respectively $H_\mathfrak{d}^\bullet$) times the exponential encoding factor $\frac{1}{r!}$ gives the coefficient of the monomial

$$mon(\mathfrak{d}) := p_{1,\lambda_1} \cdots p_{m,\lambda_m} u^r z^{1-h} q^d$$

in $\mathcal{H}_{\ell,g}$ (respectively $\mathcal{H}_{\ell,g}^{\bullet}$). If $\mathfrak{d}_1 = (h_1, d_1, r_1, \{\lambda_1, \ldots, \lambda_m\})$ and $\mathfrak{d}_2 = (h_2, d_2, r_2, \{\mu_1, \ldots, \mu_n\})$, we define

$$\mathfrak{d}_1 + \mathfrak{d}_2 := (h_1 + h_2 - 1, d_1 + d_2, r_1 + r_2, \{\lambda_1, \ldots, \lambda_m, \mu_1, \ldots, \mu_n\}). \quad (10.16)$$

Note that

$$mon(\mathfrak{d}_1 + \mathfrak{d}_2) = mon(\mathfrak{d}_1)mon(\mathfrak{d}_2),$$

which in particular shows that the addition operation we defined is associative. By $\vec{\mathfrak{d}} = (\mathfrak{d}_1^{l_1}, \ldots, \mathfrak{d}_k^{l_k})$ denote a tuple where $\mathfrak{d}_i^n$ indicates the entry $\mathfrak{d}_i$ is repeated n times. Such a tuple indexes a particular type of disconnected Hurwitz covers, as represented in Figure 10.1. Each of the parts of $\vec{\mathfrak{d}}$ gives the Hurwitz data for a connected component of the cover. We denote by $L := \sum_{j=1}^{k} l_j$ the total number of connected components of the source curve, and $|\vec{\mathfrak{d}}| = \sum_{i=1}^{k} l_i \mathfrak{d}_i$, where addition is defined in (10.16).

If you haven't keeled over yet, you may take a breath of relief now: we are done setting up notation!

Fix a monomial $mon(\mathfrak{d})$ and consider its coefficient in $\mathcal{H}_{\ell,g}^{\bullet}$: as we mentioned earlier, it is $H_{\mathfrak{d}}^{\bullet}/r!$. The disconnected Hurwitz numbers count covers that can be grouped in terms of the Hurwitz data of each connected component; hence we may express

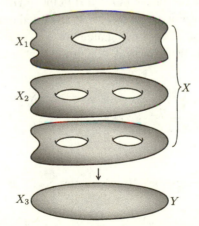

Figure 10.1 The contribution to a disconnected Hurwitz number by covers with a given number of components each supporting a ramification data. One first takes the product of connected Hurwitz numbers. If some components support identical data, one must divide by the appropriate factorial to account for overcounting and automorphisms.

$$H_{\eth}^{\bullet} = \sum_{|\vec{\eth}|=\eth} Cont(\vec{\eth}), \tag{10.17}$$

where $Cont(\vec{\eth})$ denotes the automorphism-weighted number of disconnected covers of type $\vec{\eth}$. If $\vec{\eth} = (\eth_1^{l_1}, \ldots, \eth_k^{l_k})$ as above, then we have

$$Cont(\vec{\eth}) = \binom{r}{r_1, \ldots, r_L} \frac{1}{l_1! \cdots l_k!} \prod_{j=1}^{k} (H_{\eth_j})^{l_j}, \tag{10.18}$$

where

- the quotient by the l_j!s accounts for isomorphism and automorphisms of disconnected covers corresponding to permuting the connected components with equal data;
- the multinomial coefficient accounts for all possible ways that the r additional simple ramifications can distribute themselves on the various connected components of the disconnected source curve.

Recapping what we have done so far, we have expressed a coefficient of the disconnected Hurwitz potential as follows:

$$[mon(\eth)] \mathcal{H}_g^{\bullet} = \sum_{|\vec{\eth}|=\eth} \frac{1}{r!} \binom{r}{r_1, \ldots, r_L} \frac{1}{l_1! \cdots l_k!} \prod_{j=1}^{k} (H_{\eth_j})^{l_j}. \tag{10.19}$$

Now we look at the coefficient of $mon(\mathfrak{s})$ in the Taylor expansion of $e^{\mathcal{H}_g}$. Again it can be written as a sum over $|\vec{\eth}| = \eth$, since addition of $\eth's$ corresponds to multiplication of the corresponding monomials. Carefully keeping track of the factorials in the Taylor expansion of a multivariable exponential function, one gets

$$[mon(\eth)] e^{\mathcal{H}_g} = \sum_{|\vec{\eth}|=\eth} \frac{1}{r_1! \cdots r_L!} \frac{1}{L!} \binom{L}{l_1, \ldots, l_k} \prod_{j=1}^{k} (H_{\eth_j})^{l_j}, \tag{10.20}$$

which is equal to (10.19). □

Example 10.2.2. We observe (10.15) for the coefficients of the monomial $u^4 z q^3$:

$$H_{0 \to 0,3}^{\bullet,4} \frac{u^4}{4!} z q^3 = H_{0 \to 0,3}^{4} \frac{u^4}{4!} z q^3 + \frac{1}{2!} 2 \left(H_{1 \to 0,2}^{4} \frac{u^4}{4!} q^2 \right) \left(H_{0 \to 0,1}^{0} z q \right). \tag{10.21}$$

In equation (10.21) all Hurwitz numbers don't have any λ, and to lighten notation we dropped the ($\varnothing$) from our usual notation. Equation (10.21) expresses the number of degree 3, possibly disconnected covers of $\mathbb{P}^1(\mathbb{C})$ by a rational curve, with four points of simple ramification as the sum of two terms: one corresponding to connected covers, the other to covers consisting of a disjoint union of a genus 1 curve mapping with degree 2 and a rational curve mapping isomorphically to $\mathbb{P}^1(\mathbb{C})$.

Exercise 10.2.1. Check that equation (10.21) is correct by evaluating the appropriate Hurwitz numbers.

Exercise 10.2.2. Compute the disconnected Hurwitz number $H^{\bullet,4}_{-1\to0,4} = 5$ in two ways: first using the Burnside formula (refer to Table 8.1 for the character table of S_4); then by observing that, since the genus h of the cover curve is negative, the covers must necessarily be disconnected. Hence, for any Hurwitz cover for this data, each connected component maps with degree less than or equal to 3. The corresponding connected Hurwitz numbers are therefore accessible by counting the appropriate monodromy representations. Finally organize the count of the disconnected Hurwitz number as the appropriate sum of the contributions coming from each type of disconnected cover.

Of what monomial is $H^{\bullet,4}_{-1\to0,4}/4!$ the coefficient in $\mathcal{H}^{\bullet}_\zeta$? Check that equation (10.15) holds for the coefficients of this monomial.

10.3 Cut-and-Join

The name *cut-and-join* refers to a set of recursive relations among Hurwitz numbers, obtained by analyzing what happens to the cycle type of a fixed permutation in the symmetric group S_d when composed with any simple transposition. From a geometric point of view, this is a special case of the degeneration formulas (Section 7.5), which amounts to shrinking a loop surrounding two branch points, one of which is given an arbitrary ramification profile condition, the other a simple ramification condition. Infinitely many recursions are efficiently encoded in one partial differential equation which the Hurwitz potential satisfies. This showcases the power of the generating function language to handle a large amount of combinatorial complexity, provided that the recursive structure is well tuned to the generating function encoding.

We begin this discussion by stating the elementary group theoretic fact underlying the cut-and-join recursions.

Fact 10.3.1. *Let $\sigma \in S_d$ be a fixed element of cycle type $\lambda = (n_1, \ldots, n_\ell)$, written as a composition of disjoint cycles as $\sigma = c_\ell \ldots c_1$. Let $\tau = (ij) \in S_d$*

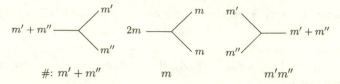

Figure 10.2 Diagrammatic illustration of cut-and-join: how a cycle is affected by
multiplication with a transposition

*vary among all transpositions. The cycle types of the composite elements $\tau\sigma$
are described below.*

> **cut.** *if i, j belong to the same cycle (say c_ℓ), then this cycle gets "cut in
> two": $\tau\sigma$ has cycle type $\lambda' = (n_1, \ldots, n_{\ell-1}, m', m'')$, with $m' + m'' =
> n_\ell$. If $m' \neq m''$, there are n_ℓ transpositions giving rise to an element of
> cycle type λ'. If $m' = m'' = n_\ell/2$, then there are $n_\ell/2$.*
>
> **join.** *if i, j belong to different cycles (say $c_{\ell-1}$ and c_ℓ), then these cycles are
> "joined": $\tau\sigma$ has cycle type $\lambda' = (n_1, \ldots, n_{\ell-1}+n_\ell)$. There are $n_{\ell-1}n_\ell$
> transpositions giving rise to an element of cycle type λ'.*

Example 10.3.2. Let $d = 4$. There are six transpositions in S_4. Pick $\sigma =
(12)(34)$ of cycle type $(2, 2)$: there are two transpositions $((12)$ and $(34))$
that "cut" σ to give rise to a transposition (cycle type $(2, 1, 1)$) and $2 \cdot 2$
transpositions $((13), (14), (23), (24))$ that "join" σ into a four-cycle.

For readers allergic to notation, Figure 10.2 presents a diagrammatic
summary of the above discussion.

Exercise 10.3.1. Prove Fact 10.3.1.

We now apply Fact 10.3.1 in the context of Hurwitz numbers. We focus our
attention on one particular branch point b on the base curve: the cut-and-join
analysis describes what happens when we shrink a loop surrounding b and a
branch point corresponding to a simple ramification.

The cut-and-join recursion involves Hurwitz numbers where the only part of
the branching that is varying is the profile of the point b. In order to lighten our
notation, we restrict our attention to a generating function for Hurwitz numbers
with only one point of non-generic ramification.

Definition 10.3.3. Setting to zero all the variables $p_{i,j}$ with $i \geq 2$ in the disconnected Hurwitz potential $\mathbf{\mathring{h}}_{\ell g}^{\bullet}$ and dropping the subscript 1 from the set of
p-variables remaining gives the potential

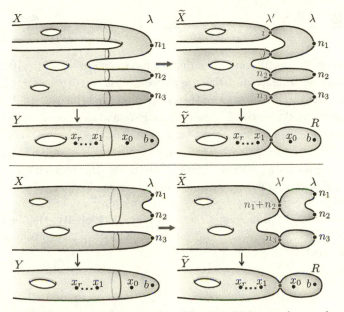

Figure 10.3 The geometric interpretation of the cut-and-join recursion as a degeneration formula. A loop surrounding b_0 and x_0 is shrunk. The type of degenerate covers occurring have a new ramification profile λ' which is obtained from λ by either cutting a cycle or joining two cycles. The cut-and-join recursion expresses this degeneration formula where the very simple Hurwitz numbers corresponding to the right-hand side of the degenerate cover are computed explicitly and hence don't appear in the formula.

$$\mathbf{S}^{\bullet}_g(p_j, u, z, q) := \sum H^{\bullet,r}_{h\to g,d}(\lambda)\, p_\lambda \frac{u^r}{r!} z^{1-h} q^d, \tag{10.22}$$

encoding Hurwitz numbers with only one special branch point. Such Hurwitz numbers are sometimes called **simple Hurwitz numbers**, and hence we call this generating function the **simple Hurwitz potential**.

The cut-and-join analysis translates to the following statement.

Theorem 10.3.4. *The disconnected simple Hurwitz potential* $\mathbf{S}^{\bullet}_g$ *is annihilated by the cut-and-join differential operator:*

$$\frac{\partial}{\partial u} - \left(\frac{1}{2} \sum_{i,j\geq 1} ijp_{i+j} z \frac{\partial^2}{\partial p_i \partial p_j} + (i+j)p_i p_j \frac{\partial}{\partial p_{i+j}} \right). \tag{10.23}$$

Proof First we observe that the statement of the theorem means that $\mathbf{S}^{\bullet}$ satisfies the following PDE:

$$\frac{\partial}{\partial u}\mathbf{S}^\bullet = \frac{1}{2}\sum_{i,j\geq 1} ijp_{i+j}z\frac{\partial^2}{\partial p_i\partial p_j}\mathbf{S}^\bullet + (i+j)p_ip_j\frac{\partial}{\partial p_{i+j}}\mathbf{S}^\bullet. \quad (10.24)$$

We analyze the coefficient of an arbitrary monomial in equation (10.24), and essentially we will recognize that it is a special case of a degeneration formula: the coefficient on the left-hand side corresponds to the Hurwitz number before the loop is shrunk; the coefficients on the right-hand side consist of the degenerate covers – the first term in the summation corresponds to covers where a cycle is cut into two cycles; the second, where two cycles are joined. Refer to Figure 10.3 for an illustration. As in the case of the degeneration formulas, the proof is cleaner with the algebraic language of monodromy representations.

Consider the monomial $p_\lambda\frac{u^r}{r!}z^{1-h}q^d$ for arbitrary values of λ, r, h and d. The coefficient of this monomial on the left-hand side of (10.24) is the Hurwitz number $H^{\bullet, r+1}_{h\to g,d}(\lambda)$, which is $1/d!$ times the number of elements in the set of monodromy representations

M^{r+1}_λ

$$= \left\{ (\sigma, \tau_0, \tau_1, \ldots, \tau_r, \alpha_1, \beta_1, \ldots, \alpha_g, \beta_g) \left| \begin{array}{l} \bullet\ \sigma \in C_\lambda, \\ \bullet\ \tau_i \text{ are simple transpositions}, \\ \bullet\ [\alpha_g, \beta_g]\ldots[\alpha_1, \beta_1]\tau_r\ldots\tau_0\sigma = e \end{array} \right. \right\}.$$

Now let Λ' denote the set of partitions $\lambda' \dashv d$ that are obtained from λ by either adding two parts of λ or splitting one of the entries of λ into two parts. For each $\lambda' \in \Lambda'$ define the set $M^r_{\lambda'}$ analogously (but note that there are now only r transpositions). Consider the natural function:

$$\Phi: M^{r+1}_\lambda \to \coprod_{\lambda'\in\Lambda'} M^r_{\lambda'},$$

defined by

$$\Phi(\sigma, \tau_0, \tau_1, \ldots, \tau_r, \alpha_1, \beta_1, \ldots, \alpha_g, \beta_g) = (\tau_0\sigma, \tau_1, \ldots, \tau_r, \alpha_1, \beta_1, \ldots, \alpha_g, \beta_g).$$

We now write the cardinality of M^{r+1}_λ as the sum of the cardinalities of the inverse images via Φ of each of the $M^r_{\lambda'}$:

$$|M^{r+1}_\lambda| = \sum_{\lambda'\in\Lambda} |\Phi^{-1}(M^r_{\lambda'})|. \quad (10.25)$$

Each of the terms on the right-hand side of (10.25) is computed making use of Fact 10.3.1. We must treat three cases separately: to slow ourselves down, we break the first case into a series of exercises.

Case 1: λ is obtained from λ' by adding two of the entries of λ': one of size i and one of size j.

Exercise 10.3.2. Making use of Fact 10.3.1 shows that for any element $x \in M_{\lambda'}^r$, the cardinality of the inverse image $\Phi^{-1}(x)$ is equal to $ijn_i n_j$, where n_i is the number of parts equal to i in λ', n_j the number of parts equal to j in λ'.

Exercise 10.3.3. Observe that the cardinality of $M_{\lambda'}^r$ is equal to $d! H_{h-1 \to g,d}^{\bullet, r}(\lambda')$.

By Exercises 10.3.2 and 10.3.3 we have the following equation:

$$|\Phi^{-1}(M_{\lambda'}^r)| = d! ijn_i n_j H_{h-1 \to g,d}^{\bullet, r}(\lambda'). \tag{10.26}$$

Exercise 10.3.4. Show that the quantity $ijn_i n_j H_{h-1 \to g,d}^{\bullet, r}(\lambda')$ is the coefficient of the monomial $p_\lambda \frac{u^r}{r!} z^{1-h} q^d$ in the expression

$$ij p_{i+j} z \frac{\partial^2}{\partial p_i \partial p_j} \mathbf{S}_g^\bullet.$$

We now consider the other two cases from Fact 10.3.1 and carry out similar analyses.

Case 2: λ is obtained from λ' by splitting a part of λ', say of size $i + j$ into one part of size i and one of size j, with $i \neq j$.

By Fact 10.3.1, for any element $x \in M_{\lambda'}^r$, the cardinality of the inverse image $\Phi^{-1}(x)$ is equal to $i + j$ times the number n_{i+j} of parts equal to $i + j$ in λ'. The cardinality of $M_{\lambda'}^r$ is equal to $d! H_{h \to g,d}^{\bullet, r}(\lambda')$, hence

$$|\Phi^{-1}(M_{\lambda'}^r)| = d!(i + j)n_{i+j} H_{h \to g,d}^{\bullet, r}(\lambda'). \tag{10.27}$$

The quantity $(i + j)n_{i+j} H_{h \to g,d}^{\bullet, r}(\lambda')$ is the coefficient of the monomial $p_\lambda \frac{u^r}{r!} z^{1-h} q^d$ in the expression

$$(i + j) p_i p_j \frac{\partial}{\partial p_{i+j}} \mathbf{S}_g^\bullet.$$

Case 3: a part of size $2i$ in λ' is split into two parts of equal size i.

In this case we must divide the RHS of (10.27) by 2, and correspondingly the term $1/2(i+i)n_{i+i} H_{h \to g,d}^{\bullet, r}(\lambda')$ is the coefficient of the monomial $p_\lambda \frac{u^r}{r!} z^{1-h} q^d$ in the expression

$$\frac{1}{2}(i + i) p_i p_i \frac{\partial}{\partial p_{i+i}} \mathbf{S}_g^\bullet.$$

By substituting (10.26), (10.27) in (10.25) and dividing the resulting equation by $d!$, we obtain the equality of the coefficients of the monomial $p_\lambda \frac{u^r}{r!} z^{1-h} q^d$ in equation (10.24): we only must notice that reorganizing the sum over $\lambda' \in \Lambda$ as a sum over all possible i, j, we must divide by $1/2$ to account for the overcounting arising from switching the roles of i and j and for the factor of $1/2$ that appears when $i = j$.   □

Exercise 10.3.5. Write down the coefficient of the monomial $p_3 \frac{u^4}{4!} z^0 q^3$ in the cut-and-join equation (10.24), and verify that the equation holds by computing directly the Hurwitz numbers involved.

The disconnected setting lends itself most naturally to formulate the cut and join recursions: return to Figure 10.3, and observe that even if the original cover $X \to Y$ is connected, after shrinking the loop the cover $\tilde{X} \to \tilde{Y}$ may be disconnected. However, when a cycle is cut, it is cut in exactly two parts. Consequently, a connected cover can be disconnected in at most two components.

One can therefore obtain a connected cut-and-join operator by adding one term corresponding to when the cut cycle disconnects the original cover. The connected cut-and-join analysis appears in Goulden and Jackson (1999). Here we state it and leave the proof as a *grand finale* exercise for our readers!

Theorem 10.3.5. *The connected simple Hurwitz potential* $\mathbf{S}_g$ *is annihilated by the connected cut-and-join operator:*

$$\frac{\partial}{\partial u} - \left(\frac{1}{2} \sum_{i,j1} ijp_{i+j} \left(z \frac{\partial^2}{\partial p_i \partial p_j} + \frac{\partial}{\partial p_i} \frac{\partial}{\partial p_j} \right) + (i+j)p_i p_j \frac{\partial}{\partial p_{i+j}} \right).$$
$$(10.28)$$

Exercise 10.3.6. Prove Theorem 10.3.5.

Appendix A
Hurwitz Theory in Positive Characteristic

Rachel Pries

A.1 Introduction

The focus of this book is on compact Riemann Surfaces, or algebraic curves defined over the complex numbers $\mathbb{C}$. There is also a beautiful story about algebraic curves defined over other fields, like the field of rational numbers $\mathbb{Q}$ or a finite field like $\mathbb{Z}/p$ where p is a prime number. In this chapter, we will give a glimpse of some interesting developments for algebraic curves defined over fields similar to the latter type, which are called fields of *positive characteristic*. In particular, we will show that the Riemann–Hurwitz formula needs to be adjusted for maps between curves in positive characteristic and that the affine line is no longer simply connected in positive characteristic.

The plan is to replace $\mathbb{C}$ by an algebraically closed field k of characteristic p and then study algebraic curves over k. But first we need to ask whether this is a valuable thing to do. One objection is that it seems as though we will lose much of the motivation for studying complex curves, such as understanding doubly-periodic complex valued functions or solving differential equations with applications to physics. An answer to this is that there are new applications involving cryptography and error-correcting codes for which algebraic curves in positive characteristic are highly useful. These include elliptic and hyperelliptic curve cryptography and Reed–Solomon and Goppa error-correcting codes, which were used by the Voyager II space mission to send pictures back to Earth.

A second objection is that it is disconcerting to work in positive characteristic (at first) because no pictures can be drawn. A lot of intuition about the genus and fundamental group of a Riemann Surface is gained by drawing loops. New theory is needed to make sense of the definitions of these objects in positive characteristic. This is true, and yet there is a strong payoff; some people would argue that these new definitions are better for studying compact Riemann Surfaces as well, because they illuminate more of the structure.

More generally, algebraic curves in positive characteristic control some of the geometry of complex curves. Studying an algebraic equation only from the perspective of the complex numbers is a lot like studying an iceberg only from above water, or studying a cubic polynomial only using its one real root. In fact, there are proofs of several theorems about complex curves which rely on information about curves in positive characteristic. This note contains a glimpse of the new phenomena and open questions about curves in positive characteristic.

A.2 Algebraic Curves in Positive Characteristic

In practice, working in positive characteristic means that all coefficients will be computed using modular arithmetic. We pick a prime number p and consider equivalence classes of numbers modulo p. This means that p acts like 0 modulo p and, more generally, $a \equiv b \bmod p$ if and only if p divides $b - a$. The set $\mathbb{Z}/p$ of equivalence classes of integers modulo p is a commutative ring, and it is a field since every nonzero element has an inverse. For example, if $p = 5$, then $2 \cdot 3 = 6 \equiv 1 \bmod 5$ and so $2^{-1} \equiv 3 \bmod 5$. One quick warning is that the exponents in equations do not reduce modulo p. For example, $2^5 = 32 \equiv 2 \bmod 5$ which does not equal $2^0 = 1 \bmod 5$. In fact, Fermat's Little Theorem says that every $a \in \mathbb{Z}/p$ is a root of the polynomial $x^p - x$ modulo p.

There is one property about $\mathbb{C}$ that is crucially important – it is algebraically closed, or, equivalently, every non-constant polynomial has a complex root. The complex number i which is a root of $x^2 + 1$ plays a key role for Riemann Surfaces. When $p = 5$ (or, more generally, when $p \equiv 1 \bmod 4$), there is still a root of $x^2 + 1$ modulo p; the number 2 acts like i algebraically because $2^2 = 4 \equiv -1 \bmod 5$. Therefore we make the convention that when we write i for a field k which is not $\mathbb{C}$, we really mean a root of the polynomial $x^2 + 1$ in k. However, it is inconvenient that there is no root of $x^2 - 2$ modulo 5. To remedy this, we add $\sqrt{2} \equiv \sqrt{-3}$ to our field, and consider the larger field of size 25 called $\mathbb{F}_{25} = \{a_0 + a_1\sqrt{2} \mid a_0, a_1 \in \mathbb{Z}/5\}$.

Without worrying about the consequences, we add the roots of all non-constant polynomials in $\mathbb{Z}/p[x]$ to make an algebraically closed field of characteristic p, which we call k. This can be made more concrete for anyone who has learned about finite fields by thinking of k as the union (or direct limit) of the finite fields $\mathbb{F}_{p^n}$ for all natural numbers n.

A.3 Smooth Algebraic Curves

An algebraic curve is an algebraic variety of dimension 1. The projective line $\mathbb{P}^1_k$ and the affine line $\mathbb{A}^1_k = \mathbb{P}^1_k - \infty$ (like Definition 2.3.1 and Remark 2.3.4) are two examples of algebraic curves defined over k. Other simple algebraic curves are affine plane curves like in Definition 3.1.2, which can be described as the set of points $(x, y) \in k^2$ satisfying a (non-degenerate) polynomial relation $F(x, y) \in k[x, y]$.

In this chapter, we focus on the affine plane curve $X^\circ_{p,t}$ defined over k given by the affine equation $F_{p,t}(x, y) = 0$, where

$$F_{p,t}(x, y) = x^p - x - y^t \qquad (A.1)$$

for a natural number t such that p does not divide t. Curves of this type are called Artin–Schreier curves and they play an important role in understanding new phenomena about maps of curves in positive characteristic.

Exercise A.3.1.

1. *Show that $X^\circ_{p,t}$ is smooth using Definition 3.1.2.*
2. *Show that the map $\phi_{p,t} : X^\circ_{p,t} \to \mathbb{A}^1_k$ given by $(x, y) \mapsto x$ is ramified only at the p points where $y = 0$, in which case it has ramification index t. (For example, when $p = 5$ and $t = 4$, then $\phi_{5,4}$ is ramified when $y = 0$ and $x \in \{0, \pm 1, \pm i\} = \{0, 1, 2, 3, 4\}$.)*
3. *Show that the map $\psi_{p,t} : X^\circ_{p,t} \to \mathbb{A}^1_k$ given by $(x, y) \mapsto y$ is not ramified anywhere.*
4. *Show that $\psi_{p,t}$ has Galois group isomorphic to $\mathbb{Z}/p$. (Hint: the automorphism $\tau(x) = x + 1$, $\tau(y) = y$ respects the equation $x^p - x - y^t$ since $(x + 1)^p \equiv x^p + 1 \bmod p$.)*

There is a unique smooth compact algebraic curve $X_{p,t}$ defined over k which contains $X^\circ_{p,t}$ as a dense open affine subset. There is one point P_∞ in $X_{p,t}$ which is not in $X^\circ_{p,t}$. In practice, it is not easy to find an equation for $X_{p,t}$ since the homogenization of $F_{p,t}$ is usually not smooth at P_∞, but here are two examples.

Exercise A.3.2.

1. *Show that the homogenization of $X^\circ_{p,t}$ in $\mathbb{P}^2_k$ is smooth at the point $[0 : 1 : 0]$ only when $t = p - 1$ or $t = p + 1$. (Hint: break the problem into the cases $t > p$ and $t < p$.)*

2. *If* $t = ap - 1$, *consider the change of variables* $\overline{y} = 1/y$ *and* $\overline{x} = x\overline{y}^a$.
   Show that the equation for $F_{p,t}$ *can be rewritten as*

$$\overline{F}_{p,t} = \overline{x}^p - \overline{x}\overline{y}^{a(p-1)} - \overline{y},$$

and show that this affine curve is smooth at the point $(\overline{x}, \overline{y}) = (0, 0)$.

A.4 The Genus and the Riemann–Hurwitz Formula

In positive characteristic, it does not make sense to define the genus of an algebraic curve as the number of its "holes". Instead, the genus is defined using differentials. Differentials are used frequently in complex analysis, where many results are proven by integrating differentials along paths and loops. A lot of the behavior of a differential is determined by its poles.

Definition A.4.1. Let X be a smooth compact algebraic curve over k. The set of differentials on X having no poles is called $\Omega(X)$. The **genus** g of X is the dimension of $\Omega(X)$ as a vector space over k.

To gain some intuition about this definition, let's look at the example of a hyperelliptic curve X with affine equation $y^2 = f(x)$ where $f(x)$ has degree R and no multiple roots. (Unlike Section 6.3, we take R to be odd so that there is a unique point P_∞ at infinity, which is also a ramification point of the double cover of the projective line.) Then $R = 2g + 1$. An example of a differential is

$$\omega_i = \frac{x^{i-1}dx}{y}.$$

Lemma A.4.2. *Let* $i \geq 1$. *On the hyperelliptic curve* $X : y^2 = f(x)$ *with* $\deg(f(x)) = 2g + 1$, *the differential* ω_i *has no poles. The set of differentials*

$$\omega_1 = \frac{dx}{y}, \quad \omega_2 = \frac{xdx}{y}, \ldots, \omega_g = \frac{x^{g-1}dx}{y}$$

is a basis for $\Omega(X)$.

Proof   Here are some of the key steps in the proof. The first step is to show that $\Omega(X)$ can be spanned by differentials of the form $x^i dx/y$. The reason is that every differential ω can be uniquely written in the form $g_1(x)dx + g_2(x)dx/y$ where g_1, g_2 are rational functions of x. Then $g_1(x)$ must be zero if ω has no pole.

The next step is to show that ω_i has no poles if and only if $1 \leq i \leq g$. To see this, we can suppose that 0 is a root of $f(x)$. For $1 \leq j \leq 2j + 1$, let $P_j = (b_j, 0)$ denote the ramification points of X other than P_∞. The function x has a double zero at $P_1 = (0, 0)$ and a pole of order 2 at P_∞. The function y has a zero of order 1 at each point P_j and (by Remark 4.3.4) a pole of order $2g + 1$ at P_∞. Note that $dx = 2ydy/f'(x)$ and $f'(x)$ is nonzero at P_j since $f(x)$ has no multiple roots. This shows that the differential dx has a zero of order 1 at each point P_j and (by Exercise 4.4.2 in Chapter 4) a pole of order 3 at P_∞.

The conclusion from this is that the differential ω_i has a zero of order $2(i - 1)$ at P_1 and a zero of order $2(g - i)$ at P_∞. In particular, ω_i has no poles if $1 \leq i \leq g$. Finally, the set $\{\omega_1, \ldots, \omega_g\}$ is linearly independent since the orders of the zero at P_∞ of these differentials are all different.   □

Here is an example about the genus of the Artin–Schreier curves from the previous section.

Proposition A.4.3. *Suppose $p \nmid t$. The curve $X_{p,t}$ with affine equation $x^p - x = y^t$ has genus $(p - 1)(t - 1)/2$.*

Is this formula believable? Let's look at the map $\phi_{p,t} : X_{p,t} \to \mathbb{P}^1$ taking $(x, y) \to x$ which has degree t. When $t = 1$, then it makes sense that $X_{p,1}$ has genus 0 because $\phi_{p,1}$ is an isomorphism. (Another way to see that $X_{p,1}$ is rational is that the function field $k(x)[y]/(x^p - x - y)$ is isomorphic to $k(x)$.) When $t = 2$ (and p is odd), then $\phi_{p,2}$ has degree 2 so $X_{p,2}$ is hyperelliptic. The genus is $g = (p-1)/2$ because the polynomial $f(x) = x^p - x$ has degree $R = p$.

One way to verify Proposition A.4.3 is to answer the following question.

Exercise A.4.1. Show that the following set of differentials is a basis for $\Omega(X_{p,t})$ and that it has cardinality $(p - 1)(t - 1)/2$:

$$\{y^b x^r dy \mid r \geq 0, b \geq 0, bp + rt \leq (p - 1)(t - 1) - 2\}.$$

Another way to verify Proposition A.4.3 is to use the Riemann–Hurwitz formula. By Exercise A.3.1, the cover $\phi_{p,t}$ is ramified at the points when $y = 0$, each with ramification index t. There are p points like this because, by Fermat's Little Theorem, the roots of $x^p - x$ modulo p are the numbers $x \in \mathbb{Z}/p$. In addition, $\phi_{p,t}$ is ramified at P_∞. This can be checked explicitly when $t = ap - 1$ using the other affine chart found in Section A.3. It can be checked more theoretically for all t by thinking about the monodromy representation from Chapter 7. At each of the $p+1$ ramification points, the differential length is $t-1$. Applying the Riemann–Hurwitz formula from Chapter 4 shows that $2g-2 = t(-2)+(p+1)(t-1)$, which gives the formula in Proposition A.4.3.

Let's see what happens for the cover $\psi_{p,t} : X_{p,t} \to \mathbb{P}^1_k$ taking (x, y) to y. In this case, the degree is p and only the point P_∞ is ramified, with ramification index p. Applying the Riemann–Hurwitz formula from Chapter 4 would give that $2g-2 = p(-2)+(p-1)$, which is clearly wrong since the genus cannot be negative.

The Riemann–Hurwitz formula as stated in Chapter 4 cannot be applied to a map of curves over k when the characteristic p divides the ramification index at a ramification point. The reason is that the differential length turns out to be greater (or even much greater) than the ramification index.

Theorem A.4.4. (Wild Riemann–Hurwitz Formula). *Suppose $f : X \to Y$ is a non-constant separable Galois cover of smooth projective curves over k. Let e_x denote the ramification index of f at a point $x \in X$.*

1. *Then*

$$2g_X - 2 = deg(f)(2g_Y - 2) + \sum_{x \in X} D_x,$$

   where the differential length D_x at x is nonzero if and only if $e_x > 1$.
2. *If p does not divide e_x, then $D_x = e_x - 1$.*
3. *If $p = e_x$, then there is a prime-to-p natural number t_x, called the ramification jump, such that $D_x = (p-1)(t_x + 1)$.*
4. *More generally, if p divides e_x, then there is a ramification filtration of the inertia group I at x which is a descending sequence of subgroups I_i of I and $D_x = \sum_{i=0}^{\infty}(|I_i| - 1)$.*

Let's try to give a final proof of Proposition A.4.3 using the wild Riemann–Hurwitz formula applied to the cover $\psi_{p,t} : X_{p,t} \to \mathbb{P}^1_k$ with equation $x^p - x = y^t$. The only ramification point is the point $x = P_\infty$ and the ramification index is $e_x = p$. By Theorem A.4.4(3), the differential length equals $(t_x + 1)(p - 1)$. To get the correct formula, we need to show that the ramification jump t_x equals t.

Now we run into the problem that we can't verify this since we haven't defined the ramification jump. To define it precisely would take too long for this note, but here is the basic idea. The cover $\psi_{p,t}$ with equation $x^p - x = y^t$ has a Galois automorphism $\tau(x) = x + 1$. At the point P_∞, we can find a uniformizer function π, which vanishes with order 1 at P_∞, and so the valuation val(π) is 1 in the local ring of the curve at P_∞. Then the ramification jump is defined to be $t_{P_\infty} = \text{val}(\tau(\pi) - \pi) - 1$.

Exercise A.4.2. Let $t = ap - 1$. From Exercise A.3.2(2), remember the change of variables $\overline{y} = 1/y$ and $\overline{x} = x\overline{y}^a$ which let us rewrite the equation for $F_{p,t}$ as $\overline{F}_{p,t} = \overline{x}^p - \overline{x}\overline{y}^{a(p-1)} - \overline{y}$. The point P_∞ is the point $(\overline{x}, \overline{y}) = (0, 0)$.

1. Check that a uniformizer at P_∞ is $\pi = \overline{x}$.
2. Check that $\tau(\pi) - \pi = \overline{y}^a$.
3. Check that $\text{val}(\overline{y}^a) = ap$ and that $t_{P_\infty} = ap - 1 = t$.

A.5 The Affine Line Is Not Simply Connected Over k

The complex plane is simply connected; every loop in $\mathbb{C}$ can be retracted to a point and so is homotopy equivalent to the identity. This means that the fundamental group of $\mathbb{C}$ is trivial. Using Section 5.3, this is equivalent to saying that there are no unramified covers of $\mathbb{C}$ having degree greater than 1. (Such a cover would also give a cover of the projective line ramified only above ∞, which would contradict the Riemann–Hurwitz formula from Chapter 4.)

All of the facts in the previous paragraph are false in positive characteristic. The analogue of the complex plane in positive characteristic is the affine line $\mathbb{A}_k^1$. By Exercise A.3.1(3), the map $\psi_{p,t} : X_{p,t}^\circ \to \mathbb{A}^1$ given by $(x, y) \mapsto y$ is not ramified anywhere.

The fundamental group of the affine line is defined as the inverse limit of the Galois groups of all the unramified Galois covers of the affine line. Its degree p quotients correspond to $\mathbb{Z}/p$-Galois unramified covers of the affine line, which in turn are described by Artin–Schreier extensions of the form $x^p - x = f(y)$ where $f(y)$ is a polynomial. By studying Artin–Schreier curves, we can see that the fundamental group $\pi^1(\mathbb{A}_k^1)$ of the affine line over k is nontrivial, and also that it is truly huge (an infinitely generated profinite group). This is because the degree p quotients of the fundamental group are indexed both by the discrete invariant of the degree of $f(y)$ and by the parameters of the coefficients of $f(y)$. By Abhyankar's Conjecture (proven by Harbater and Raynaud), it is known that every finite group generated by elements of p-power order is a quotient of $\pi^1(\mathbb{A}_k^1)$, but there is currently no conjecture about the structure of $\pi^1(\mathbb{A}_k^1)$.

Here are some good references if you would like to learn more about these topics: Hirschfeld, Korchmáros, and Torres (2008) and Stichtenoth (2009).

References

Hirschfeld, J. W. P., G. Korchmáros and F. Torres. (2008). *Algebraic Curves Over a Finite Field*. Princeton Series in Applied Mathematics. Princeton, NJ: Princeton University Press, pp. xx+696. ISBN: 978-0-691-09679-7.

Stichtenoth, Henning. (2009). *Algebraic Function Fields and Codes*. Second ed. Vol. 254. Graduate Texts in Mathematics. Berlin: Springer-Verlag, pp. xiv+355. ISBN: 978-3-540-76877-7.

Appendix B
Tropical Hurwitz Numbers

Hannah Markwig and Dhruv Ranganathan

B.1 Tropical Geometry: Where Does It Come From?

At its heart, algebraic geometry can be viewed as the study of the geometry of solutions to systems of polynomial equations. For the moment, let us work with polynomials in n variables over a field k, so our variety X is affine, and it naturally lives in k^n. When k has a norm, such as $\mathbb{C}$ with its usual Euclidean norm, one can look at the *coordinate-wise sizes* of the points of X. Concretely, for any point $(x_1, \ldots, x_n) \in X \subset k^n$, consider $(|x_1|, \ldots, |x_n|) \in \mathbb{R}^n_{\geq 0}$.

The variety X may now be studied in two steps: (1) study all possible sizes of solutions in $\mathbb{R}^n_{\geq 0}$ and (2) study all the solutions that have a fixed size.

One might wonder why the set of coordinate-wise sizes has any reasonable structure whatsoever, and why studying it is at all useful. After all, we have just replaced an algebraic object with something that is quite non-algebraic: inequalities are built into the very definition of a norm. In fact, that such objects have a nice structure is not immediately clear, and this set of sizes of solutions must be modified slightly before the structure is visible. Fix a real number b, a base for a logarithm, and study the map

$$\tau_b : \mathbb{C}^n \to (\mathbb{R} \sqcup \{-\infty\})^n,$$

$$(z_1, \ldots, z_n) \mapsto (\log_b |z_1|, \ldots, \log_b |z_n|).$$

The image of X under τ_b is known as the *amoeba* of X. Define the *tropicalization* of X to be $\operatorname{trop}(X) = \lim_{b \to 0} \tau_b(X)$. This brings us to the first remarkable property of the tropicalization.

Theorem B.1.1. *Let X be a connected subvariety of $\mathbb{C}^n$ of complex dimension d. The tropicalization of X is a polyhedral complex of real dimension equal to d.*

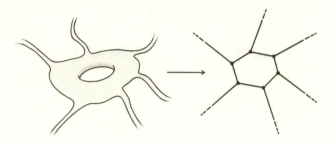

Figure B.1   A punctured genus 1 curve and its tropicalization

A polyhedral complex is an object glued from polyhedra in a nice fashion. The tropicalization has a nice combinatorial and piecewise-linear structure, while also remembering at least one important piece of information about the variety X: its dimension. In particular, an algebraic curve in $\mathbb{C}^n$ gives rise to a graph, such as the picture in Figure B.1. The industry of tropical geometry is largely about determining exactly what is remembered in this process, and exploiting the simplicity of tropicalizations to understand properties and invariants of X itself. For further details on this perspective on the subject, see for instance (Gathmann, 2006; Maclagan and Sturmfels, 2015).

B.2 Axiomatic Tropical Geometry

While tropical techniques are often used to study classical objects, tropical geometric objects have a rich and complicated geometry of their own. Another point of view on the subject is to study the geometry of polynomial functions on the real numbers where the operations of addition and multiplications are redefined. Precisely, the **max-plus semifield** is $(\mathbb{R} \cup \{-\infty\}, \oplus, \odot)$, where

$$a \oplus b := \max\{a, b\} \quad \text{and} \quad a \odot b := a + b.$$

The additive identity element in this semifield is $-\infty$, the multiplicative identity is 0 and there are no additive inverses. A tropical polynomial $\bigoplus_{i=0}^{n} a_i \odot x^{\odot i}$ (where the notation $x^{\odot i}$ stands for the i-th power of x, taken with respect to the tropical multiplication, i.e. $x^{\odot i} := x \odot \ldots \odot x$) is a piecewise linear function $\max\{a_i + i \cdot x\}$. The same holds true for multivariate polynomials. Recall how polynomials over $\mathbb{C}$ are used to define algebraic curves, i.e. Riemann Surfaces that come embedded into $\mathbb{C}^2$ (Definition 3.1.2). A tropical polynomial similarly gives a tropical curve embedded in $(\mathbb{R} \cup \{-\infty\})^2$, defined as the corner locus of the tropical polynomial, or the locus where the corresponding function fails to be locally linear. Embedded tropical curves are piecewise linear graphs satisfying a so-called balancing condition. As is the case for Riemann

Surfaces, it is often easier to study tropical curves and their covers without a particular choice of embedding, and we will swiftly switch to this viewpoint here. To learn more about embedded tropical curves, Richter-Gebert, Sturmfels and Theobald (2003), Gathmann (2006) and Maclagan and Sturmfels (2015) are excellent starting points.

In this chapter, we axiomatically introduce the tropical analogue of the double Hurwitz problem – covers of $\mathbb{P}^1(\mathbb{C})$ with ramification profiles μ over 0, ν over ∞, and only simple ramification otherwise. While tropical covers can be defined in more general settings (Bertrand, Brugallé and Mikhalkin, 2014; Cavalieri, Markwig and Ranganathan, in press), the double Hurwitz case is particularly well suited to the context of this book. Using the monodromy data of the cover, one obtains a correspondence theorem, which equates a weighted count of tropical covers with the solution to the given Hurwitz problem. This allows one to use tropical techniques to study algebraic properties of Hurwitz numbers. We conclude with a discussion of how this axiomatic definition arises directly from algebraic geometry via a process known as *degeneration*.

B.3 Tropical Covers

We now introduce tropical covers of the tropical line. Just as the projective line is formed from the affine line by adding a single point, the tropical model of $\mathbb{P}^1(\mathbb{C})$ is formed from $\mathbb{R} \sqcup \{\infty\}$ by adding a single point: that is, $\mathbb{P}^1_{trop} := \mathbb{R} \sqcup \{\pm\infty\}$.

An **(abstract, explicit) tropical curve** is a connected metric graph Γ satisfying the following properties. A vertex is called a **leaf** if it is one-valent and an (inner) vertex otherwise. An edge e is called an **end** and has length $l(e) = \infty$ if it is adjacent to a leaf; otherwise it is called a **bounded edge** and has a length $l(e) \in \mathbb{R}$. The **valence** $val(V)$ of each (inner) vertex is at least 3.

The **genus** of the tropical curve Γ is defined to be the genus of the underlying graph Γ, which is (see (5.6))

$$g(\Gamma) = E - V + 1.$$

Here, E and V are the numbers of edges and vertices of Γ, respectively.

Next we define the notion of a good map of tropical curves.

Definition B.3.1. A **tropical cover** of $\mathbb{R} \cup \{\pm\infty\}$ is a tuple (Γ, f), where Γ is a tropical curve and $f : \Gamma \rightarrow \mathbb{R} \cup \{\pm\infty\}$ is a continuous map satisfying:

- f sends only one-valent vertices to $\{\pm\infty\}$. We refer to the ends adjacent to vertices mapping to $-\infty$ as **left ends**, and to the ones whose adjacent vertices map to $+\infty$ as **right ends**.

- If we orient the edges of Γ according to the order of the images of their vertices in $\mathbb{R} \cup \{\pm\infty\}$ and consider f restricted to an edge e as a map from the interval $[0, l(e)]$ to its image (where we identify the left endpoint of e with 0 and the right endpoint with $l(e)$ and use x for a coordinate on the segment), f is integer affine linear, which means:

$$f(x) = f(0) + w(e)x,$$

with $f(0) \in \mathbb{R}$ and $w(e) \in \mathbb{N}$. The stretching factor $w(e)$ is referred to as the **weight** of an edge e.
- At each vertex, the **balancing condition** holds: the sum of the weights of the incoming edges equals the sum of the weights of the outgoing edges.

The balancing condition implies a well-defined notion of **degree**, namely the weighted number of preimages of a generically chosen point in $\mathbb{R}$.

Example B.3.2. Figure B.2 shows a tropical cover of genus 1 and degree 4. The numbers appearing next to edges are the weights. We do not specify edge lengths in the picture, since they are implicitly given by their image interval and the weight.

Two covers $f : \Gamma \to \mathbb{R} \cup \{\pm\infty\}$ and $f' : \Gamma' \to \mathbb{R} \cup \{\pm\infty\}$ are called isomorphic if there is an isomorphism φ of the underlying abstract tropical curves (i.e. a homeomorphism respecting the edge lengths) satisfying $f' \circ \varphi = f$.

Remark B.3.3. Assume $f : \Gamma \to \mathbb{R} \cup \{\pm\infty\}$ is a cover such that Γ has only one-valent and three-valent vertices, and such that the images of the three-valent vertices are distinct. Then the only automorphisms arise due to **wieners** and **balanced forks** as in (Cavalieri, Johnson and Markwig, 2010,

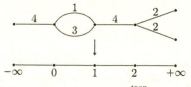

Figure B.2 A tropical cover contributing to $H^{trop}_{1 \to 0}((4), (2, 2))$ as in Definition B.3.4

Figure B.3 Local pictures of a wiener and a balanced fork

Lemma 4.2); see Figure B.3. The automorphism group of such a cover thus has size $|\mathrm{Aut}(f)| = 2^{W+B}$, where W denotes the number of wieners and B denotes the number of balanced forks.

We are now ready to define tropical double Hurwitz numbers:

Definition B.3.4. Fix two partitions μ and ν of an integer $d \geq 1$ and a genus g, such that $r := 2g - 2 + \ell(\mu) + \ell(\nu) > 0$. Fix r pairwise distinct points $p_1, \ldots, p_r$ in $\mathbb{R}$. The **tropical Hurwitz number** $H^{trop}_{d \atop g \to 0}(\mu, \nu)$ is defined as the weighted number $H^{trop}_{d \atop g \to 0}(\mu, \nu) = \sum_f m(f)$ of tropical degree d covers $f : \Gamma \to \mathbb{R} \cup \{\pm\infty\}$ where

- Γ is a tropical curve of genus g;
- the tuple of weights of left ends is μ and the tuple of weights of right ends is ν;
- the preimage $f^{-1}(p_i)$ contains a vertex of Γ.

The multiplicity $m(f)$ with which a cover $f : \Gamma \to \mathbb{R} \cup \{\pm\infty\}$ contributes to $H^{trop}_{d \atop g \to 0}(\mu, \nu)$ is defined as

$$m(f) = \frac{1}{|\mathrm{Aut}(f)|} \cdot \prod_e w(e),$$

where the product goes over all bounded edges e of Γ.

Keeping track of the Euler characteristic of Γ and applying the tropical version of the Riemann–Hurwitz formula, one can show that Γ has only trivalent inner vertices, and that there are $r > 0$ such vertices. By Remark B.3.3, the factor $\frac{1}{|\mathrm{Aut}(f)|}$ then equals $\frac{1}{2^{B+W}}$, where W denotes the number of wieners and B the number of balanced forks. The definition of $m(f)$ is motivated by tropical intersection theory of tropical moduli spaces of covers (Cavalieri, Johnson and Markwig, 2010). It is straightforward to see that the definition does not depend on the choice of the points $p_1, \ldots, p_r$; for simplicity we always assume $p_1 = 0, \ldots, p_r = r - 1$, as in the cover in Figure B.2. With this convention fixed, the combinatorial type of the source curve (i.e. the tropical curve without the data of the metric) together with the weights of all edges define the data of a tropical cover.

Example B.3.5. Figure B.4 demonstrates the tropical count of $H^{trop}_{4 \atop 1 \to 0}((4), (2,2))$. Besides the cover appearing in Figure B.2 which contributes $6 = \frac{1}{2} \cdot 1 \cdot 3 \cdot 4$, the three depicted covers contribute 4, 3 and 1 respectively, leading to a total of $H^{trop}_{4 \atop 1 \to 0}((4), (2, 2)) = 14$.

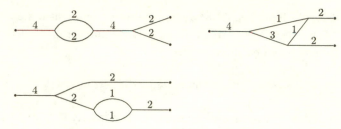

Figure B.4 The covers contributing to $H^{trop}_{\substack{4\\1\to0}}((4),(2,2))$, in addition to the one depicted in Figure B.2

B.4 Tropical Covers as Shadows – Movies of Monodromy Representations

Why are tropical covers shadows of holomorphic maps between Riemann Surfaces? Here, we explain this by taking a detour via the symmetric group, in terms of monodromy representations as in Definition 7.1.6. More generally, tropical covers can be viewed as a graphical organization of the degeneration formulas in Theorem 7.5.1. The main consequence of the interpretation of tropical covers as shadows of holomorphic maps is the fact that tropical and usual Hurwitz numbers agree:

Theorem B.4.1 (Correspondence Theorem, [Cavalieri, Johnson and Markwig, 2010]). *Fix two partitions μ and ν of an integer $d \geq 1$ and a genus g such that $0 < r = 2g - 2 + \ell(\mu) + \ell(\nu)$. Then the Hurwitz number $H_{\substack{d\\g\to0}}(\mu,\nu)$ from Definitions 6.1.6 (resp. 10.1.5) agrees with the tropical Hurwitz number $H^{trop}_{\substack{d\\g\to0}}(\mu,\nu)$ from Definition B.3.4.*

By Theorem 7.3.1, the Hurwitz number $H_{\substack{d\\g\to0}}(\mu,\nu)$ equals the number of monodromy representations of the appropriate type times $\frac{1}{d!}$. To prove Theorem B.4.1, one has to show that each monodromy representation gives rise to a tropical cover, and that $\frac{1}{d!}$ times the number of monodromy representations giving rise to the tropical cover $f : \Gamma \to \mathbb{R} \cup \{\pm\infty\}$ equals the multiplicity $m(f)$ with which f contributes to $H^{trop}_{\substack{d\\g\to0}}(\mu,\nu)$. To give more details, we demonstrate how a monodromy representation gives rise to a tropical cover. The claim about the multiplicity follows after a careful analysis of the cut-and-join relations in this case (see Fact 10.3.1). The connectedness condition for a monodromy representation translates to the connectedness of the source of the tropical cover.

Recall that the relevant connected monodromy representations of covers of $\mathbb{P}^1(\mathbb{C})$ are tuples $(\sigma_1, \tau_1, \ldots, \tau_r, \sigma_2)$ of elements of S_d such that the subgroup generated by the σ_i and τ_i acts transitively on $\{1, \ldots, d\}$, such that σ_1 has cycle type μ, τ_i is a transposition for all i, σ_2 has cycle type ν and the product $\sigma_2 \circ \tau_r \circ \ldots \circ \tau_1 \circ \sigma_1 = e$ (see Example 7.1.8 and Theorem 7.3.1).

Given a monodromy representation $(\sigma_1, \tau_1, \ldots, \tau_r, \sigma_2)$, the associated tropical cover can simply be viewed as a movie depicting the cycle types of σ_1, $\tau_1 \circ \sigma_1$, $\tau_2 \circ \tau_1 \circ \sigma_1$, $\ldots$, $\tau_r \circ \ldots \circ \tau_1 \circ \sigma_1 = \sigma_2^{-1}$; these cycle types are represented by a collection of appropriately weighted edges over the intervals $(-\infty, 0], [0, 1], [1, 2], \ldots, [r-2, r-1], [r-1, \infty) \subset \mathbb{R}$. The real line can be interpreted as the time of our movie: the way it is divided into intervals divides the movie in time sections, each corresponding to one permutation from the list above. By Fact 10.3.1, the multiplication with a transposition either cuts an edge of weight m into two edges of weights m_1 and m_2, such that $m = m_1 + m_2$ or joins two edges of weights m_1 and m_2 to one edge of weight $m = m_1 + m_2$. If we connect the weighted edges we draw above each interval accordingly, we thus obtain (the combinatorial type of) a tropical.cover – the balancing condition is satisfied.

Example B.4.2. The tuple $((1234), (12), (12), (13), (12)(34))$ contributes to the Hurwitz number $H^{trop}_{\substack{4 \\ 1 \to 0}}((4), (2, 2))$: the subgroup generated by these elements clearly acts transitively on $\{1, \ldots, 4\}$, the cycle types are as required, and the product of all the entries is $e \in S_4$. To depict our movie for this monodromy representation, we draw an edge of weight 4 over $(-\infty, 0]$ (representing the permutation (1234)), dividing into two edges of weights 1 and 3 over $[0, 1]$ (representing the permutation $(12) \circ (1234) = (1)(234)$), joining back to an edge of weight 4 over $[1, 2]$ (again representing (1234)) and finally dividing into two edges of weight 2 each over $[2, \infty)$ (representing $(13) \circ (1234) = (12)(34)$). In other words, we draw the tropical cover from Figure B.2.

The correspondence theorem opens up the possibility to study Hurwitz numbers in terms of tropical covers.

This point of view was very useful in exploring the structure of double Hurwitz numbers: Goulden, Jackson and Vakil (2005), observed that double Hurwitz numbers $H_{\substack{d \\ g \to 0}}(\mu, \nu)$ are piecewise polynomial functions of the entries of the partitions μ and ν.

Some interesting properties of such functions were proved using tropical covers and the correspondence theorem in Cavalieri, Johnson and Markwig (2011).

B.5 Bending and Breaking

In the last section, we defined tropical covers in analogy with algebraic covers, then used representation theory to deduce their relationship to the geometric objects that we started with. In this final section, we make a direct connection between algebraic and tropical curves, and scratch the surface of a beautiful theory of degenerations.

B.5.1 Breaking Curves: Degenerations

We have seen that Riemann Surfaces can be cut out by homogeneous polynomial equations with coefficients in $\mathbb{C}$, but one could just as easily work with polynomial equations whose coefficients lie in the ring $\mathbb{C}[t]$ – polynomials in many variables, whose coefficients are polynomials in one variable. The reader may wish to think of t as being a "time index", though one might find the idea of complex- and negative-valued time revolting. For each time index t, one obtains a single Riemann Surface as before. Degeneration is what that happens at special values of t, when one Riemann Surface breaks into multiple Riemann Surfaces, glued together; see Figure B.5.

Consider the polynomial $f_t = xy - tz^2$. For each time index t_0 that is nonzero, the vanishing locus of f_{t_0} in $\mathbb{P}^2(\mathbb{C})$ is a smooth conic curve, i.e. it is isomorphic to $\mathbb{P}^1$. However, when $t = 0$, you can check that the vanishing locus of f_0 is the union of two coordinate axes. We denote $\mathscr{X}_t$ the collection of all solutions of the polynomial f_t for all possible values of t. For each specific value of t the solution to $f_t = 0$ is a Riemann Surface which we denote X_t. We call $\mathscr{X}_t$ a **family** of Riemann Surfaces[1] and X_t the fiber of the family at time t.

For a slightly more complicated example, consider the family $\mathscr{X}_t$ defined by the polynomial $g_t = t \cdot (x^3 + y^3 + z^3) - xyz$. For each time index t that is nonzero, this produces a smooth curve X_t in the complex projective plane, having genus 1. However, X_0 is the union of three lines. Notice the following phenomenon: *the special member of this family was broken into multiple pieces, but each piece was geometrically simpler.* This is the principle of degeneration techniques: one attempts to extract information about a

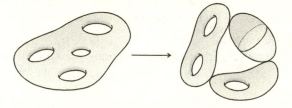

Figure B.5 A degeneration of a genus 4 curve into three pieces

---

[1] There are certain niceness conditions that one must place on such a family to avoid pathologies, but we ignore this for our discussion.

curve by first breaking it into many simpler pieces. However, as a *conservation of complexity law*, the simpler the pieces, the more complex their interaction. In a moment, we will see a concrete manifestation of this.

B.5.2 Dual Graphs and Degenerations

Suppose one has a family of Riemann Surfaces $\mathscr{X}_t$, as above. Suppose that for $t \neq 0$, X_t is a Riemann Surface, and suppose that X_0 is a nodal Riemann Surface. Informally, nodal means that X_0 is a finite collection of Riemann Surfaces, glued to each other at points, such that no three meet at a single point. We now define a graph Γ associated to $\mathscr{X}_t$, which we call the **dual graph** of the family: declare the set of vertices to be the set of irreducible components of the nodal curve X_0 (the various smooth "pieces" of the nodal Riemann Surface). For each point at which two such pieces meet, place an edge between the corresponding vertices.

Let us denote the pieces of X_0 as $C_1, \ldots, C_r$. Then we have the formula, for any $t_0 \neq 0$,

$$g(X_0) = \sum_{i=1}^{r} g(C_i) + g(\Gamma),$$

where $g(\Gamma)$ denotes the genus of the graph Γ. This is an illustration of the conservation of complexity principle above. When the pieces of X_0 are the simplest possible, i.e. each piece is a sphere, then $g(C_i) = 0$ and $g(\Gamma) = g$. The number $g(\Gamma)$ measures how far the graph Γ is from being a tree, and we see that, for such degenerations, *all the complexity has been transferred to the graph*. There is an interesting special case: the surface X_0 consists only of spheres, where each sphere is glued to other spheres at exactly three points. By applying an appropriate Möbius transformation to each sphere, one can always move these three gluing points to 0, 1 and ∞. In this case, something amazing has happened: *the tropical object Γ completely determines the classical object X_0*. We call this a **maximal degeneration**.

Even in the nicest cases, the dual graph alone forgets a lot of information: in general there are distinct nodal curves that have the same dual graph. Or, there can exist two families $\mathscr{X}_t$ and $\mathscr{Y}_t$ such that X_0 and Y_0 coincide, but at other time indices, the families are different. It is useful to work with graphs that have a little more information. To do this, we add lengths to the edges of the graph, turning it into a metric graph.

Let us choose an edge e and focus our attention locally on the node of X_0, corresponding to e. Using the local equations that cut out $\mathscr{X}_t$ one can measure how fast[2] this node forms. For instance, contrast the family of curves given by $xy = t^2 z^2$ with the one given by $xy = tz^2$. Near $t = 0$, the first node forms twice as fast as the second. This process can be made precise to give lengths to each edge in the graph.

This process produces, from each family of degenerating Riemann Surfaces, a tropical curve. One can build maps between families of algebraic curves, just as one builds maps between the curves themselves. When the degenerations are chosen to be a maximal degeneration, all the complexity of the family of covers is transferred to the complexity of graphs, and one obtains a geometric manifestation of the tropical covers in the previous section. This gives us a geometric link between families of Hurwitz covers of Riemann Surfaces and tropical covers.

References

Bertrand, Benoît, Erwan Brugallé and Grigory Mikhalkin. (2014). "Genus characteristic numbers of tropical projective plane". *Compos. Math.* 150.**1**. arXiv:1105.2004, pp. 46–104.

Cavalieri, Renzo, Paul Johnson and Hannah Markwig. (2010). "Tropical Hurwitz numbers". *J. Algebr. Comb.* 32.**2**. arXiv:0804.0579, pp. 241–265. DOI: 10.1007/s10801-009-0213-0.

Cavalieri, Renzo, Paul Johnson and Hannah Markwig. (2011). "Wall crossings for Double Hurwitz numbers". *Adv. Math.* 228.**4**. arXiv:1003.1805, pp. 1894–1937.

Cavalieri, Renzo, Hannah Markwig and Dhruv Ranganathan. "Tropicalizing the space of admissible covers". Preprint, arXiv:1401.4626.

Gathmann, Andreas. (2006). "Tropical algebraic geometry". *Jahresbericht der DMV* 108.**1**. arXiv:math.AG/0601322, pp. 3–32.

Goulden, Ian, David M. Jackson and Ravi Vakil. (2005). "Towards the geometry of double Hurwitz numbers". *Adv. Math.* 198, pp. 43–92.

Maclagan, Diane and Bernd Sturmfels. (2015). *Introduction to Tropical Geometry.* Vol. 161. Graduate Studies in Mathematics. AMS.

Richter-Gebert, Jürgen, Bernd Sturmfels and Thorsten Theobald. (2003). "First steps in tropical geometry". *Idempotent Mathematics and Mathematical Physics, Proceedings Vienna.* arXiv:math/0306366.

---

[2] We can't be too general here, but in the case of an equation of the form $xy = f(t)$ we define the *speed that the node forms at* as the order of vanishing of the function $f(t)$ at 0, or alternatively, the exponent of the first nonzero coefficient in the Taylor expansion of f at $t = 0$.

Appendix C
Hurwitz Spaces

Paul Johnson

As a topological space, the surface of genus g is unique – that is, if X and Y are two genus g surfaces, there is a homeomorphism $f : X \to Y$. However, if we give X and Y complex structures and think about them as Riemann Surfaces, they no longer have to be "the same" – we can't necessarily turn the homeomorphism f into a biholomorphic map.

Understanding the different ways we can view a genus g topological surface as a Riemann Surface is a fundamental (if vague) question in geometry. It turns out that for $g = 0$ there is a unique way to put a complex structure on a sphere – this is why it makes sense to talk about *the* Riemann Sphere. However, for $g > 0$, there are infinitely many distinct complex structures. Thus, counting the number of complex structures is not an interesting problem, and we need a more precise meaning of what it means to "understand" them. The answer lies in the concept of a *moduli space*.

C.1 Moduli Spaces

A moduli space $\mathcal{M}$ is, first of all, a topological space. What makes a moduli space more than just a topological space is that the underlying set of points is naturally in bijection with some interesting geometric objects. A bit more precisely, each point of $\mathcal{M}$ corresponds to an isomorphism class of some object we want to study, and the intuitive idea is that two points in $\mathcal{M}$ are "close" to each other if the corresponding isomorphism classes of geometric objects are "close" to one another. Let us now look at some examples to refine our understanding.

Example C.1.1 (Projective space $\mathbb{CP}^n$). In Definition 2.3.1, $\mathbb{P}^n(\mathbb{R})$ was introduced as the set of lines in $\mathbb{R}^{n+1}$ through the origin. Another way of saying this is that $\mathbb{P}^n$ is "the moduli space of lines through the origin in $\mathbb{R}^{n+1}$". Each point of the moduli space corresponds to a line, and moving continuously from a point to a nearby point amounts to continuously wiggling the corresponding lines.

Example C.1.2 (Moduli space of curves $\mathcal{M}_g$). Similarly, we have the moduli space of genus g curves $\mathcal{M}_g$. Each point in $\mathcal{M}_g$ corresponds to an isomorphism class of genus g Riemann Surfaces. That is, we can view any genus g Riemann Surface X_g as a point of $\mathcal{M}_g$. Two Riemann Surfaces X_g and Y_g represent the same point of $\mathcal{M}_g$ if they are isomorphic (Definition 4.1.3). The topology on $\mathcal{M}_g$ is complicated to describe formally, but satisfies some intuitive properties. For instance, if X is given by the vanishing of a homogeneous polynomial $P(X, Y, Z)$, then changing the coefficients of P by a little bit will result in a new Riemann Surface Y that we think of as "near" X.

Example C.1.3 (Moduli space of elliptic curves $\mathcal{M}_1$). An **elliptic curve** is another name for a Riemann Surface of genus 1, which were discussed in Section 3.2.2. There, we were told that every genus 1 curve could be represented as a complex torus $\mathbb{C}/\Lambda$, and furthermore that we could take the lattice Λ to be generated by the vectors 1 and τ.

This description of genus 1 Riemann Surfaces gives us a way to understand the topology on $\mathcal{M}_1$. To find the curves "close" to a given curve X, realize X as $\mathbb{C}/\langle 1, \tau \rangle$ and then change the complex number τ slightly.

This suggests that $\mathcal{M}_1$ is a particularly nice topological space – each neighborhood of a point seems to be isomorphic to $\mathbb{C}$, and so $\mathcal{M}_1$ looks like a Riemann Surface itself!

This winds up being not quite correct, because some lattices, and hence some genus 1 curves, have extra symmetries. Consider the "square" lattice, where $\tau = i$: this lattice has an extra symmetry of rotation by ninety degrees. If we deform this lattice slightly to a new lattice Λ_ϵ by changing τ to $(1 + \varepsilon)i$, our torus is rectangular, with the imaginary side slightly longer than the other real side. Similarly, if we shrink τ slightly to $(1 - \delta)i$, we get a rectangular lattice Λ_δ, but now the real direction is longer than the imaginary.

But rotating and scaling the lattice doesn't change the resulting Riemann Surface, and so these should really be the same Riemann Surface. More precisely, consider the map from $\mathbb{C}$ to $\mathbb{C}$ given by multiplication by $(1 - \delta)i$. This sends the generator 1 of Λ_ϵ to $(1-\delta)i$, a generator of Λ_δ. It sends the generator $(1 + \epsilon)i$ of Λ_ϵ to $-(1 + \epsilon)(1 - \delta)$. We would like this to be a generator of Λ_δ, namely -1, and some algebra shows that if $\delta = \epsilon/(1+\epsilon)$, this in fact the case. Thus, we see that multiplication by $(1 - \delta)i$ gives a biholomorphism between the Riemann Surfaces $\mathbb{C}/\Lambda_\epsilon$ and $\mathbb{C}/\Lambda_\delta$, and so these two points represent the same point of $\mathcal{M}_1$.

Geometrically, it can be thought that the point on the moduli space that corresponds to the square lattice has 180 degrees around it, instead of 360, like a point at the edge of a piece of paper. The two edges of the piece of paper have been identified together to make a cone point.

Our vague question about "understanding" the complex structures on a genus g surface is now less vague – we want to "understand" the space $\mathcal{M}_g$. That may still sound vague, but now it is clear how to pose precise questions – anything we can ask about topological spaces, we can ask about $\mathcal{M}_g$. In particular: Is $\mathcal{M}_g$ a manifold? If so, what dimension does it have? Is $\mathcal{M}_g$ connected? What is $\pi_1(\mathcal{M}_g)$? Hurwitz theory arose as a means of answering some of these questions.

C.2 Hurwitz Spaces

Hurwitz spaces are moduli spaces of ramified covers. To define the Hurwitz number $H_{h \xrightarrow{d} g}(\lambda_1, \ldots, \lambda_n)$, it was necessary to fix distinct points of ramification $b_1, \ldots, b_n \in Y$ in order to get finitely many covers. Hurwitz spaces are what we get if we allow the points b_i to move.

Definition C.2.1. Fix a Riemann Surface Y of genus g, integers d and h, and n partitions $\lambda_1, \ldots, \lambda_n$ of d. The **Hurwitz space** $\mathcal{H}_{h \xrightarrow{d} Y}(\lambda_1, \ldots, \lambda_n)$ is the set of all Hurwitz covers of Y, as in Definition 6.1.6 of the Hurwitz numbers, except that now $(b_1, \ldots, b_n) \in Y$ is any set of n distinct points in Y.

We have described the underlying set of a Hurwitz space. What about the topology? Once we fix the b_is, there are only finitely many covers. So intuitively, deforming a cover slightly amounts to moving the b_is slightly. If we don't move the b_is too far, we can keep each permutation in the monodromy representation the same and still have a Hurwitz cover.

An exciting feature of moduli spaces: since points of a moduli space correspond to geometric objects, one can describe functions between moduli spaces by describing corresponding operations on the geometric objects themselves. We can see an example of this using projective spaces: assigning to each line through the origin in $\mathbb{R}^3$ its vertical projection onto the xy-plane defines a function from (an open set of) $\mathbb{P}^2(\mathbb{R})$ to $\mathbb{P}^1(\mathbb{R})$. When a map between moduli spaces is defined by forgetting some of the geometric structure of the objects in the first moduli space (such as the map just described, where we forgot all information about the vertical component in the direction of lines), it is commonly called a *forgetful map*.

Back to our Hurwitz spaces, one of the reasons $\mathcal{H}_{h \xrightarrow{d} Y}(\lambda_1, \ldots, \lambda_n)$ is so useful is because it has two natural forgetful maps to other spaces.

The branch map. There is a map br that forgets everything but the branch points. More formally, let

$$\Delta_Y \subset Y^n = \{(b_1, \ldots, b_n) | b_i = b_j \text{ for some } i, j\}$$

so that $D(Y, n) = Y^n \setminus \Delta_Y$ is the moduli space of n distinct points on Y. Then

$$br : \mathcal{H}_{h \xrightarrow{d} Y}(\lambda_1, \ldots, \lambda_n) \to D(Y, n)$$

sends a Hurwitz cover to its set $b_1, \ldots, b_n$ of branch points.

The source map. The map s that forgets the "cover of Y" part and just remembers the source curve X:

$$s : \mathcal{H}_{h \xrightarrow{d} Y}(\lambda_1, \ldots, \lambda_n) \to \mathcal{M}_g.$$

The space $\mathcal{M}_g$ is a complicated but important space, while the space $D(Y, n)$ is a rather simple space. If we understand the map br, that will let us understand the Hurwitz space, and then if we understand the map s we can transfer this to information about $\mathcal{M}_g$.

Theorem C.2.2. *The map $br : \mathcal{H}_{h \xrightarrow{d} Y}(\lambda_1, \ldots, \lambda_n) \to D(Y, n)$ is a covering map.*

Proof We give only the main idea. The idea behind br being a local homeomorphism is that the only way to deform a Hurwitz cover is by deforming the points of ramification b_i; but this is exactly deforming a point in $D(Y, n)$. To show the covering property at a point $(b_1, \ldots, b_n)$, take contractible open neighborhoods U_i around b_i that are pairwise disjoint; then $U_1 \times \cdots \times U_n$ is an open set in $D(Y, n)$ whose inverse image consists of homeomorphic disjoint open sets above. $\square$

Note that the degree of this covering map is the Hurwitz number $H_{h \xrightarrow{d} g}(\lambda_1, \ldots, \lambda_n)$, where g is the genus of Y.

C.3 Applications of Hurwitz Spaces

Applications of Hurwitz spaces typically involve all or mostly simple ramification. In particular, we will use $\mathcal{H}_{d,r}$ as shorthand for the Hurwitz space of degree d covers of $\mathbb{CP}^1$, with r points of simple ramification. By the Riemann–Hurwitz formula, we see that X has genus $r/2 + 1 - d$. Alternatively, if we fix g, then $r = 2g - 2 + 2d$.

C.3.1 The Dimension of $\mathcal{M}_g$

The first application of Hurwitz spaces was Riemann's calculation of the dimension of $\mathcal{M}_g$ in the middle of the nineteenth century.

Theorem C.3.1. $\dim_{\mathbb{C}} \mathcal{M}_g = 3g - 3$.

Proof We do not give a complete proof, but merely sketch how the theorem uses Hurwitz spaces.

First, observe that $D(\mathbb{P}^1(\mathbb{C}), r)$ is a manifold with complex dimension r: a point in $D(\mathbb{P}^1(\mathbb{C}), r)$ consists of a configuration of r points $(b_1, \ldots, b_r)$ on $\mathbb{P}^1(\mathbb{C})$. Each of the b_is can move independently from all others in $\mathbb{P}^1(\mathbb{C})$, a one-dimensional manifold; hence you have r degrees of freedom in moving around $D(\mathbb{P}^1(\mathbb{C}), r)$. Since br is a covering map, then $\mathcal{H}_{d,r}$ is also a manifold of complex dimension r. Thus, if we could show that $s : \mathcal{H}_{d,r} \rightarrow \mathcal{M}_g$ was surjective, and could find the dimension of a generic fiber F_s of s, then we would know the dimension of $\mathcal{M}_g$:

$$\dim \mathcal{M}_g = r - \dim(F_s). \tag{C.1}$$

This is exactly the argument Riemann made.

Proofs for either of these facts require more algebraic geometry than we can reproduce here. It turns out that s is surjective whenever $d > 2g$. In this case the generic fiber of s has dimension $2d - g + 1$. Since the dimension of $\mathcal{H}_{d,r}$ is $r = 2g - 2 + 2d$, we see by (C.1) the dimension of $\mathcal{M}_g$ must be $(2g - 2 + 2d) - (2d - g + 1) = 3g - 3$. $\square$

If you are paying attention, you should be upset with this result – you have been told that there is a unique complex structure on the sphere, but if we plug in $g = 0$ the theorem, we get -3. Similarly, the discussion of genus 1 curves suggests that $\dim \mathcal{M}_1 = 1$, while Riemann's result tells us that the dimension should be 0.

We were imprecise in stating Theorem C.3.1. Riemann Surfaces of genus 0 and 1, which have Euler characteristic 2 and 0, respectively, have very different behavior from Riemann Surfaces of genus at least 2 when the Euler characteristic is negative.

As an example, it is a result of Hurwitz that Riemann Surfaces of genus at least 2 have finite automorphism groups. This is not true for genus 0 and genus 1 surfaces. In Remark 4.5.3, you were told that the group of mobius transformations $\mathbb{P}GL(2, \mathbb{C})$, which has complex dimension 3, acts as automorphisms on $\mathbb{CP}^1$. Similarly, any genus 1 curve is $\mathbb{C}/\Lambda$, and since $\mathbb{C}$ is a group under addition and Λ is a subgroup, $\mathbb{C}/\Lambda$ is a group and acts on itself by translations. Thus $\mathbb{C}/\Lambda$ has automorphism group of dimension 1.

The correct general statement is that $\dim \mathcal{M}_g - \dim \mathrm{Aut}(C_g) = 3g - 3$ for any g.

C.3.2 The Moduli Space of Curves Is Connected

Another application of Hurwitz spaces, due first to Klein at the end of the nineteenth century, is to show that $\mathcal{M}_g$ is connected. As before, we use the forgetful maps. We know that $s : \mathcal{H}_{d,r} \to \mathcal{M}_g$ is surjective for d large, and the surjective image of a connected space is connected. Thus, it is enough to show that $\mathcal{H}_{d,r}$ is connected for d large.

To do this, we use the fact that $br : \mathcal{H}_{d,r} \to D(\mathbb{P}^1(\mathbb{C}), r)$ is a covering map. It is clear that $D(\mathbb{P}^1(\mathbb{C}), r)$ is a connected space (if it's not clear, convince yourself that this just amounts to being able to move continuously one configuration of r points to any other configuration), so to show that $\mathcal{H}_{d,r}$ is connected it is enough to pick one point $\mathbf{b} = (b_1, \ldots, b_r) \in D(\mathbb{P}^1(\mathbb{C}), r)$ in the base space, and show that all the points $br^{-1}(\mathbf{b})$ lie in the same component. But since br is a covering map, checking this is the same as checking whether the fundamental group of $D(\mathbb{P}^1(\mathbb{C}), r)$ acts transitively on $br^{-1}(\mathbf{b})$. (Note that the argument is the same as in Exercise 7.1.6, even if the base space is not a Riemann Surface.)

There is something amusing and potentially confusing going on here. Since in $br^{-1}(\mathbf{b})$ we have fixed the ramification points, the set $br^{-1}(\mathbf{b})$ consists of usual Hurwitz covers. Thus, each point of $br^{-1}(\mathbf{b})$ corresponds to a monodromy representation. However, we now have a *different* group $\pi_1(D(\mathbb{P}^1(\mathbb{C}), r))$ acting on these monodromy representations, and this action is also called monodromy. The group $\pi_1(D(\mathbb{P}^1(\mathbb{C}), r))$ is called a *surface braid group* and it is well understood. In particular, it has an explicit presentation, where generators are given by moving two consecutive ramification points b_i and b_{i+1} in a circle around each other until they come back to their original positions. Spend a moment and meditate on why we just described a loop in $D(\mathbb{P}^1(\mathbb{C}), r)$.

The action of these generators on the monodromy representations corresponding to Hurwitz covers can be explicitly written out (this is not hard – try it!), and be seen to act transitively (this is more taxing).

C.3.3 ELSV Formula

To end, we briefly mention an exciting application of Hurwitz numbers. The ELSV formula, from the end of the twentieth century, and named for its discovers, Ekedhal, Lando, Shapiro and Vainshtein, gives a formula for certain integrals over the moduli space of curves in terms of Hurwitz numbers. The integrals involved are fundamentally important but relatively intractable, while we have seen that Hurwitz numbers are combinatorial and computable. This connection is a key step in Okounkov and Pandharipande's proof of Witten's conjecture, which states that generating functions for a family of integrals over the moduli spaces of curves satisfy the same systems of differential equations that model the propagation of waves in shallow water.

The ELSV formula has several other applications and it is responsible for much of the current interest in Hurwitz numbers.

C.3.4 Further Reading

A friendly but precise introduction to the concept of moduli space is contained in Chapter 0 of Kock and Vainsencher (2007). Riemann's argument for the dimension of the moduli space of curves is reviewed with more detail in Vakil's expository paper (Vakil, 2008). Klein's original argument for the connectedness of the moduli space of curves is found in Klein (1963). The ELSV formula appears first in Ekedahl et al. (2001). Okounkov and Pandharipande's proof of the Witten conjecture is in Okounkov and Pandharipande (2009). For a friendly graduate-level textbook on the moduli spaces of curves, Harris and Morrison (1998) is always a good read.

References

Ekedahl, Torsten et al. (2001). "Hurwitz numbers and intersections on moduli spaces of curves". *Invent. Math.* 146, pp. 297–327.

Harris, Joseph and Ian Morrison. (1998). *Moduli of Curves*. Springer.

Klein, Felix. (1963). *On Riemann's Theory of Algebraic Functions and Their Integrals. A Supplement to The Usual Treatises*. Translated from the German by Frances Hardcastle. New York: Dover Publications, Inc., pp. xii+76.

Kock, Joachim and Israel Vainsencher. (2007). *An Invitation to Quantum Cohomology*. Vol. 249. Progress in Mathematics. Kontsevich's formula for rational plane curves. Boston, MA: Birkhäuser Boston Inc., pp. xiv+159.

Okounkov, A. and R. Pandharipande. (2009). "Gromov–Witten theory, Hurwitz numbers, and matrix models". In: *Algebraic Geometry—Seattle 2005. Part 1*. Vol. 80.

Proc. Sympos. Pure Math. Providence, RI: Amer. Math. Soc., pp. 325–414. URL: http://dx.doi.org/10.1090/pspum/080.1/2483941.

Vakil, R. (2008). "The moduli space of curves and Gromov–Witten theory". In: *Enumerative Invariants in Algebraic Geometry and String Theory*. Vol. 1947. Lecture Notes in Math. Berlin: Springer, pp. 143–198. URL: http://dx.doi.org/10.1007/978-3-540-79814-9_4.

Appendix D
Does Physics Have Anything to Say About Hurwitz Numbers?

Vincent Bouchard

What? Physics? Why would physics have anything to do with Hurwitz numbers? Interesting question, isn't it?

Well, it turns out that physics – in particular, string theory – *does* indeed have much to say about Hurwitz numbers, and enumerative geometry in general. In this appendix I will try to explain why physics has deep connections with enumerative invariants such as Hurwitz numbers. I will not be precise; nor will I state explicit results or theorems (in fact, there is not a single equation in this appendix!). Rather, my goal is simply to convey some of the fascinating ideas behind the connection between string theory and enumerative geometry. Hopefully, by the end of the appendix, you will find these relations interesting enough to delve into the literature, where you can find precise conjectures and theorems!

D.1 Physical Mathematics

For many physicists, mathematics is seen as a tool: a language for building models of nature. However, in the last forty years or so, a fascinating new research area has flourished: using physics as a tool to further our understanding of mathematics. To some pure mathematicians, this statement may sound like an abomination. But let me try to convince you that physics indeed has much more to say about mathematics itself than one may expect.

While this interconnection between physics and mathematics is certainly not new (historically, physics and mathematics have always been intimately related), it has been very successful in recent years. So successful that it has been given its own name: *Physical Mathematics* (Moore, 2014). In the particular case where the corner of physics studied is string theory, it is also sometimes called *String-Math*[1]. The idea is simple but far-reaching: use the complex structural properties of physical theories to discover new connections between different areas of mathematics.

One of the most useful tools that physicists have at their disposal is physical dualities. Roughly speaking, physicists are interested in constructing mathematical models that explain the universe (and provide predictions that can be tested in experiments, following the scientific method). But sometimes, it happens that more than one mathematical model provides the same observable quantities (or, at least, observables that are in a one-to-one relationship, and/or perhaps only in some appropriate limit). When this is the case, from a physics standpoint both models are valid physical descriptions. We say that these models are *dual*.

But dualities have an unforeseen consequence. Dual physical models may be constructed using completely different mathematical structures. For instance, a given physical model may be formulated in the language of algebraic geometry, while a dual model may involve objects in number theory or topology. Then, the physical duality "implies" a connection between certain objects (the observables) defined in *a priori* disconnected areas of mathematics! And more often than not, these connections are rather unexpected, and would have been difficult to guess by mathematicians without prior knowledge of the physical duality.

As it turns out, such connections are often very useful, beyond simply relating different areas of mathematics. It may happen that some quantities that are difficult to calculate in one model become easy to compute in the dual model; or that certain quantities in one model, when grouped together in a certain way, must satisfy startling mathematical properties, because of their interpretation in the dual model. Countless examples of stunning new mathematical results have been obtained in recent years by exploiting physical dualities in mathematics.

The drawback, however, from a mathematical viewpoint, is that physical dualities are usually not rigorously proved. Thus, generally speaking, physical dualities give rise to conjectures, but rarely do they actually produce theorems right away. Once the conjecture is out, it may take years for mathematicians to actually prove (or disprove!) it.

But this is precisely the main use of physical dualities in mathematics: as a bottomless pool of (often far-reaching) ideas and conjectures for mathematicians. After all, in mathematics, our goal is to prove theorems, but first and

---

[1] String-Math is an annual series of international conferences that started in 2011 at the University of Pennsylvania, bringing together mathematicians and physicists working on mathematical ideas related to string theory.

foremost we must come up with good ideas for statements that we want to prove. Physical dualities are an endless source of ideas to play with.

D.2 Hurwitz Numbers and String Theory

You are probably thinking, sure, this is all good, but what does it have to do with Hurwitz numbers? So keeping physical mathematics in mind, let me now switch gears and try to explain a connection between Hurwitz numbers and physics.

D.2.1 String Theory

One of the fundamental questions in theoretical physics is whether there exists a unified mathematical model for quantum physics and gravity. On the one hand, quantum field theory provides a successful mathematical model for the three basic fundamental interactions underlying the Standard Model of particle physics. On the other hand, Einstein's general relativity describes the gravitational interaction with impressive accuracy. One attempt at unifying these two theories is to apply the methods of quantum field theory to general relativity, but it turns out that gravity is "non-renormalizable", which basically means that this naive approach fails at high energies. Something else is at play.

Perhaps the most promising attempt at unifying quantum physics and gravity (simultaneously providing a quantum theory of gravity) is string theory. The fundamental idea behind string theory is fairly simple.

In ordinary particle physics, we model fundamental particles mathematically as points: that is, 0-dimensional objects, moving in space. Moreover, we usually combine the three dimensions of space and the time dimension into a single four-dimensional entity (or manifold, mathematically), known as spacetime. Thus, a point moving in space traces a real line in spacetime, which we call the worldline of a particle. Naively, particle physics may be understood as a theory of real lines in a four-dimensional manifold.

String theory postulates that fundamental particles are not points, but rather one-dimensional extended objects: strings. If the length of the strings is very small, those would be indistinguishable from point-like particles at the energies reached in current experiments. There are two types of strings: loops (closed strings) and line segments (open strings). Each type moves in space, and as such traces real surfaces in spacetime. String theory may then be understood as a theory of real surfaces in a manifold.

With this being said, it would be naive to think that we can delve into the details of string theory in a short appendix. Let me instead focus on a few salient properties of string theory.

First, there are different flavors of string theory. It turns out that many of these distinct string models are related by physical dualities, known as string dualities. As such, going back to the discussion above, string theory is a perfect playground for physical mathematics!

Second, in the models of string theory that are most promising for describing our universe, it appears that spacetime is 10-dimensional. It is usually modeled as the Cartesian product of 4-dimensional Minkowski spacetime and a 6-dimensional compact Riemannian manifold (the so-called "extra dimensions"). Roughly speaking, topological and geometrical properties of this 6-dimensional compact manifold dictate the low-energy physics that results from the string model in our observable Minkowski spacetime.

But mathematically, string theory is very complicated. In fact, much remains to be discovered about the fundamental mathematical structure underlying string theory. Nevertheless, one could try to apply the ideas of physical mathematics to string dualities to conjecture new connections in mathematics. To this end, it is often more successful to play with "toy models" of string theory instead of the more realistic string models. These toy models are still interesting physically – they compute certain sectors of the fully-fledged string models – but more importantly for us, they generally lie on strong and well-established mathematical foundations.

D.2.2 Topological String Theory

One such simplified version of string theory is called *closed A-model topological string theory*. In this model, 10-dimensional spacetime is replaced by a compact 6-real-dimensional manifold X, which is assumed to be Kähler. In this case, the string theory becomes a theory of holomorphic curves in X. In fact, this is not quite precise; rather, the theory is really a theory of holomorphic (stable) maps from Riemann Surfaces to the target space X. The observables of the theory only depend on the Kähler structure of X, and not on its complex structure. In fact, they have a very precise definition: they correspond mathematically to generating functions for certain rational numbers that somehow "count" pseudoholomorphic curves in X. More precisely, these rational numbers are obtained by taking integrals of certain cohomology classes over the moduli space of stable maps from compact Riemann Surfaces with a certain number of marked points to X. They are called Gromov–Witten invariants of X, and play a fundamental role in enumerative geometry.

A-model topological string theory can also be generalized to include open strings. The theory then becomes a theory of holomorphic (stable) maps from bordered Riemann Surfaces (Riemann Surfaces with boundaries) to X; the boundaries of the Riemann Surfaces must map to a fixed Lagrangian submanifold[2] $L \subset X$, called a *brane*. For particular choices of (X, L), such as X being a toric Calabi–Yau manifold (or orbifold) and L belonging to a particular class of Lagrangian submanifolds with topology $\mathbb{R}^2 \times S^1$, known as toric branes, open A-model topological string theory can be defined rigorously mathematically, via a procedure called localization (Katz and Liu, 2002). The observables then generate so-called open Gromov–Witten invariants of (X, L).

There is also another version of topological string theory: the B-model. It is also a theory of maps from Riemann Surfaces to a 6-real-dimensional manifold Y, but it is much simpler. In fact, the observables of the theory only depend on the complex structure Y and not on its Kähler structure.

A particularly striking example of a string duality, then, is the relation between the A-model and the B-model. This duality, known as *mirror symmetry*, states that the A-model on a target space X is dual to the B-model on a different target space Y. The duality somehow "identifies" – in a highly nontrivial way – the Kähler structure of X with the complex structure of its mirror manifold Y (Hori et al., 2003). A remarkable consequence of mirror symmetry is that generating functions of Gromov–Witten invariants of X can be rewritten in terms of classical (and much simpler to evaluate) integrals on the mirror side. That's a particularly compelling example of physical mathematics at play!

D.2.3 The Connection with Hurwitz Numbers

But you are probably still wondering: what does any of this have to do with Hurwitz numbers?

At least we are getting closer. On the one hand, what we now know is that A-model topological string theory is, mathematically, a theory of maps from Riemann Surfaces to a target space X. On the other hand, Hurwitz numbers involve ramified coverings from Riemann Surfaces to the complex projective line $\mathbb{P}^1$. Both problems involve maps from Riemann Surfaces to a target space. Could they be related?

It turns out that they are indeed related, and in many different ways. Here I will only highlight one of the connections between string theory,

---

[2] A Lagrangian submanifold is a submanifold whose dimension is half of X and over which the restriction of the Kähler form vanishes.

Gromov–Witten invariants and Hurwitz numbers, but there are many other very interesting relations.

We have seen that open A-model topological string theory can be defined rigorously mathematically for target spaces (X, L) with X a toric Calabi–Yau manifold or orbifold and L a toric brane, giving rise to open Gromov–Witten invariants. In fact, to be more precise, it turns out that open Gromov–Witten invariants depend on one more piece of data: an integer $f \in \mathbb{Z}$ called the framing of the brane $L \subset X$. The framing dependence arises in the application of the localization procedure to define open Gromov–Witten invariants.

The simplest example of a toric Calabi–Yau threefold is $X = \mathbb{C}^3$. Generating functions for open Gromov–Witten invariants can be calculated explicitly in this setup, and it turns out that they are intimately related to simple Hurwitz numbers counting ramified coverings of $\mathbb{P}^1$ with arbitrary ramification over $\infty \in \mathbb{P}^1$ and simple ramification elsewhere. Indeed, perhaps miraculously, it turns out that if we take the $f \to \infty$ limit of the open Gromov–Witten generating functions for $\mathbb{C}^3$, we obtain precisely generating functions for simple Hurwitz numbers (Bouchard and Mariño, 2008; Caporaso et al., 2007)!

The most direct proof of this connection involves a rewriting of both open Gromov–Witten invariants and simple Hurwitz numbers. On the one hand, as is explained in Appendix C, simple Hurwitz numbers can be rewritten in terms of Hodge integrals: that is, integrals over the moduli space of curves with marked points – this is called the ELSV formula (Ekedahl et al., 2001). On the other hand, open Gromov–Witten invariants of $\mathbb{C}^3$ can also be rewritten in terms of integrals over the moduli space of curves with marked points, albeit more complicated integrals. This is known as the topological vertex formalism (Aganagic et al., 2005; Li et al., 2009; Maulik et al., 2011). With these reformulations established, one can calculate explicitly the $f \to \infty$ limit to prove the connection between these enumerative invariants (Bouchard and Mariño, 2008).

A direct consequence of establishing a connection between Hurwitz numbers and Gromov–Witten invariants is that we can now take advantage of string dualities to study mathematical properties of Hurwitz numbers. Mirror symmetry, in particular, leads to unexpected new results in Hurwitz theory. For instance, it has been conjectured (Bouchard et al., 2009; Mariño, 2008) (and proved very recently in Eynard and Orantin (2012) and Fang, Melissa Liu and Zong (2013)) that for the open A-model on (X, L) with X a toric Calabi–Yau manifold or orbifold and L a toric brane, the mirror B-model has a very simple and explicit formulation in terms of a topological recursion that originated in the context of matrix models (Chekhov, Eynard and Orantin, 2006; Eynard and Orantin, 2007). The recursive description is formulated in terms of complex analysis of a complex curve, the spectral curve, that somehow encapsulates the geometric data characterizing X and L. In other words, all open Gromov–Witten invariants of (X, L) can be reconstructed from simple complex analysis of the spectral curve. Quite remarkable.

But then, by evaluating appropriately the $f \to \infty$ limit on the mirror B-model side, it follows that generating functions of simple Hurwitz numbers must also satisfy a topological recursion for a specific spectral curve! The spectral curve can be calculated explicitly via the $f \to \infty$ limit, and it turns out that the spectral curve corresponding to simple Hurwitz numbers is the Lambert curve $x = ye^{-y}$ – so called because of its similarity to the Lambert W function, related to the enumeration of trees in combinatorics, that satisfies $z = W(z)e^{W(z)}$. What we just found is that all simple Hurwitz numbers can be reconstructed in a very explicit way, via a topological recursion, from the Lambert curve (Bouchard and Mariño, 2008)!

To be fair, physical mathematics only produces a conjecture that all simple Hurwitz numbers can be reconstructed recursively from the Lambert curve. But by now several proofs have been formulated, independently from the string theory interpretation. Perhaps the most direct approach involves the ELSV formula and the topological vertex formalism; in this context, what can be shown is that the cut-and-join equation satisfied by simple Hurwitz numbers (Goulden and Jackson, 1997) can be recast into the topological recursion for the Lambert curve (Eynard, Mulase and Safnuk, 2011). But the connection is highly nontrivial. Alternative proofs have been obtained by rewriting generating functions of simple Hurwitz numbers as matrix integrals (Borot et al., 2011), or by exploiting the polynomiality property of simple Hurwitz numbers (Dunin-Barkowski et al., 2015a).

It should be noted that this relation between Hurwitz numbers and string theory can be generalized beyond simple Hurwitz numbers, for instance to double Hurwitz numbers counting ramified coverings of $\mathbb{P}^1$ with arbitrary ramification over $0, \infty \in \mathbb{P}^1$ and simple ramification elsewhere. On the one hand, a formula analogous to the ELSV formula exists for double Hurwitz numbers, but the Hodge integrals must be replaced by Hurwitz–Hodge integrals over the moduli space of stables maps from twisted Riemann Surfaces with marked points to the classifying spaces $\mathcal{B}\mathbb{Z}_a$ (Johnson, Pandharipande and Tseng, 2011). On the other hand, one can consider the open A-model on the toric orbifolds $X = [\mathbb{C}^3/\mathbb{Z}_a]$. Open Gromov–Witten invariants can then also be rewritten in terms of Hurwitz–Hodge integrals (Brini and Cavalieri, 2011; Ross, 2011; Ross and Zong, 2013). It can be shown that the two theories are again related via the $f \to \infty$ limit (Bouchard et al., 2014).

Mirror symmetry applied to the orbifolds $X = \left[\mathbb{C}^3 / \mathbb{Z}_a \right]$ then implies that a certain class of double Hurwitz numbers (so-called orbifold Hurwitz numbers) can also be obtained via the topological recursion, but with spectral curve now given by $x^a = ye^{-ay}$. A proof of this statement can be obtained along similar lines as for simple Hurwitz numbers, using the reformulations in terms of Hurwitz–Hodge integrals and the cut-and-join equation for double Hurwitz numbers (Bouchard et al., 2014; Do, Leigh and Norbury, 2012). Alternatively, the connection can also be proven using quasi-polynomiality (Cavalieri, Johnson and Markwig, 2010) of double Hurwitz numbers (Dunin-Barkowski et al., 2015b).

It is possible to generalize this connection to include all double Hurwitz numbers. However, it is unclear how to proceed for more general Hurwitz numbers. This is certainly an interesting avenue to investigate further!

It should finally be mentioned that the connection between Hurwitz numbers and string theory gives rise to many other fascinating results. One of them is the existence of a "quantum curve" for Hurwitz numbers (Bouchard et al., 2014; Mulase, Shadrin and Spitz, 2013). Unfortunately, limited space prevents me from elaborating further on these interesting developments, but if you find the notion of "quantum curves" intriguing you are certainly encouraged to delve deeper into the accompanying literature!

D.2.4 Physical Mathematics and Enumerative Geometry

Physical mathematics is a relatively new area of research, but it has already led to far-reaching new results in various areas of mathematics. String theory in particular is an exciting playground for physical mathematics, due to the ubiquitous presence of string dualities. Mirror symmetry is an example of such duality with striking implications for enumerative invariants. In this short appendix I could only touch very superficially upon the subject. But I hope that I have managed to convey the excitement surrounding physical mathematics and its application to enumerative geometry and Hurwitz theory in particular. There are undoubtedly many more fascinating new results to come!

References

Aganagic, Mina et al. (2005). "The topological vertex". *Communications in Mathematical Physics* 254, pp. 425–478. arXiv: hep-th/0305132 [hep-th].

Brini, Andrea and Renzo Cavalieri. (2011). "Open orbifold Gromov–Witten invariants of [$\mathbb{C}^3/\mathbb{Z}_n$]: localization and mirror symmetry". *Selecta Mathematica* 17, pp. 879–933. arXiv: 1007.0934 [math.AG].

Borot, Gaetan et al. (2011). "A matrix model for simple Hurwitz numbers, and topological recursion". *Journal of Geometry and Physics* 62.**2**, pp. 522–540. arXiv: 0906.1206 [math-ph].

Bouchard, Vincent and Marcos Mariño. (2008). "Hurwitz numbers, matrix models and enumerative geometry". Vol. 78. Proceedings of Symposia in Pure Mathematics. AMS, pp. 263–283. arXiv: 0709.1458 [math.AG].

Bouchard, Vincent et al. (2009). "Remodeling the B-model". *Communications in Mathematical Physics* 287, pp. 117–178. DOI: 10.1007/s00220-008-0620-4. arXiv: 0709.1453 [hep-th].

Bouchard, Vincent et al. (2014). "Mirror symmetry for orbifold Hurwitz numbers". *Journal of Differential Geometry* 98.**3**, pp. 375–423. arXiv: 1301.4871 [math.AG].

Caporaso, Nicola et al. (2007). "Phase transitions, double-scaling limit and topological strings". *Physical Reviews* D 75, p. 046004. arXiv: hep-th/0606120 [hep-th].

Cavalieri, Renzo, Paul Johnson and Hannah Markwig. (2010). "Chamber Structure of Double Hurwitz Numbers". arXiv: 1003.1805 [math.AG].

Chekhov, Leonid, Bertrand Eynard and Nicolas Orantin. (2006). "Free energy topological expansion for the 2-matrix model". *JHEP 0612*, p. 053. arXiv: math-ph/0603003 [math-ph].208

Do, Norman, Oliver Leigh and Paul Norbury. (2012). "Orbifold Hurwitz numbers and Eynard–Orantin invariants". arXiv: 1212.6850 [math.AG].

Dunin-Barkowski, Petr et al. (2015a). "Polynomiality of Hurwitz numbers, Bouchard–Mariño conjecture, and a new proof of the ELSV formula". *Advances in Mathematics* 279, pp. 67–103. arXiv: 1307.4729 [math.AG]

Dunin-Barkowski, Petr et al. (2015b). "Polynomiality of orbifold Hurwitz numbers, spectral curve and a new proof of the Johnson–Pandharipande–Tseng formula". arXiv: 1504.07440 [math-ph].

Ekedahl, T. et al. (2001). "Hurwitz numbers and intersections on moduli spaces of curves". *Inventiones Mathematicae* 146, pp. 297–327. arXiv: math/0004096 [math.AG].

Eynard, Bertrand, Motohico Mulase and Brad Safnuk. (2011). "The Laplace transform of the cut-and-join equation and the Bouchard–Mariño conjecture on Hurwitz numbers". *Publications of the Research Institute for the Mathematical Sciences* 47, pp. 629–670. arXiv: 0907.5224 [math.AG].

Eynard, Bertrand and Nicolas Orantin. (2007). "Invariants of algebraic curves and topological expansion". *Communications in Number Theory and Physics* 1, pp. 347–452. arXiv: math-ph/0702045 [math-ph].

Eynard, Bertrand and Nicolas Orantin. (2012). "Computation of open Gromov–Witten invariants for toric Calabi–Yau 3-folds by topo-logical recursion, a proof of the BKMP conjecture". arXiv: 1205.1103 [math-ph].

Fang, Bohan, Chiu-Chu Melissa Liu and Zhengyu Zong. (2013). "All genus open-closed mirror symmetry for affine toric Calabi-Yau 3-orbifolds". arXiv: 1310.4818 [math.AG].

Goulden, Ian P. and David M. Jackson. (1997). "Transitive factorisations into transpositions and holomorphic mappings on the sphere". *Proceedings of the American Mathematical Society* 125, pp. 51–60.

Hori, Kentaro et al. (2003). *Mirror Symmetry*. Vol. 1. Clay Mathematics Monographs. American Mathematical Society.

Johnson, Paul, Rahul Pandharipande and Hsian-Hua Tseng. (2011). "Abelian Hurwitz–Hodge integrals". *Michigan Mathematical Journal* 60, pp. 171–198. arXiv: 0803.0499 [math.AG].

Katz, Sheldon and Chiu-Chu Melissa Liu. (2002). "Enumerative geometry of stable maps with Lagrangian boundary conditions and multiple covers of the disk". *Advances in Theoretical and Mathematical Physics* 5, pp. 1–49. arXiv: math/0103074 [math.AG].

Li, Jun et al. (2009). "A mathematical theory of the topological vertex". *Geometry and Topology* 13, pp. 527–621. arXiv: math/0408426 [math.AG].

Mariño, Marcos. (2008). "Open string amplitudes and large order behavior in topological string". *JHEP* 0803, p. 060. arXiv: hep-th/0612127 [hep-th].

Maulik, Davesh et al. (2011). "Gromov–Witten/Donaldson–Thomas correspondence for toric 3-folds". *Inventiones Mathematicae* 186, pp. 435–479. arXiv: 0809.3976 [math.AG].

Moore, Greg. (2014). *Physical Mathematics and the Future*. Summary talk given at the Strings conference, Princeton. URL: http://www.physics.rutgers.edu/~gmoore/PhysicalMathematicsAndFuture.pdf.

Mulase, Motohico, Sergey Shadrin and Loek Spitz. (2013). "The spectral curve and the Schrödinger equation of double Hurwitz numbers and higher spin structures". In: *Communications in Number Theory and Physics* 7.1, pp. 125–143. arXiv: 1301.5580 [math.AG].

Ross, Dustin. (2011). "Localization and gluing of orbifold amplitudes: the Gromov–Witten orbifold vertex". *Transactions of the American Mathematical Society* 366, p. 3. arXiv: 1109.5995 [math.AG].

Ross, Dustin and Zhengyu Zong. (2013). "The gerby Gopakumar–Mariño–Vafa formula". *Geometry and Topology* 17, pp. 2935–2976. arXiv: 1208.4342 [math.AG].

Bibliography

Armstrong, Mark Anthony. (1983). *Basic Topology*. Undergraduate Texts in Mathematics. Corrected reprint of the 1979 original. New York–Berlin: Springer-Verlag, pp. xii+251. ISBN: 0-387-90839-0.

Axler, Sheldon. (1997). *Linear Algebra Done Right*. 2nd ed. Undergraduate Texts in Mathematics. New York: Springer-Verlag, pp. xvi+251. ISBN: 0-387-98258-2. DOI: 10.1007/b97662. URL: http://dx.doi.org/10.1007/b97662.

Burnside, W. (1955). *Theory of Groups of Finite Order*. 2nd ed. New York, NY: Dover Publications, Inc., pp. xxiv+512.

Caporaso, L. (2014). "Gonality of algebraic curves and graphs". In *Algebraic and Complex Geometry*. Vol. 71. Springer Proc. Math. Stat. Springer, Cham, pp. 77–108. DOI: 10.1007/978-3-319-05404-9_4. URL: http://dx.doi.org/10.1007/978-3-319-05404-9_4.

Conway, John B. (1978). *Functions of One Complex Variable*. 2nd ed. Vol. 11, Graduate Texts in Mathematics. New York–Berlin: Springer-Verlag, pp. xiii+317. ISBN: 0-387-90328-3.

Dummit, David S. and Richard M. Foote. (2004). *Abstract Algebra*. 3rd ed. Hoboken, NJ: John Wiley & Sons, Inc., pp. xii+932. ISBN: 0-471-43334-9.

Fulton, William and Joe Harris. (1991). *Representation Theory*. Vol. 129, Graduate Texts in Mathematics. A First Course, Readings in Mathematics. New York: Springer-Verlag, pp. xvi+551. ISBN: 0-387-97527-6; 0-387-97495-4. DOI: 10.1007/978-1-4612-0979-9. URL: http://dx.doi.org/10.1007/978-1-4612-0979-9.

Goulden, I. P. and D. M. Jackson. (1999). "A proof of a conjecture for the number of ramified coverings of the sphere by the torus". In *J. Combin. Theory Ser. A* 88.2 pp. 246–258. ISBN: 0097-3165. DOI: 10.1006/jcta.1999.2992. URL: http://dx.doi.org/10.1006/jcta.1999.2992.

Hatcher, Allen. (2002). *Algebraic Topology*. Cambridge University Press, pp. xii+544. ISBN: 0-521-79160-X; 0-521-79540-0.

Hilton, P. J. and U. Stammbach. (1997). *A Course in Homological Algebra*. 2nd ed. Vol. 4, Graduate Texts in Mathematics. New York: Springer-Verlag, pp. xii+364. ISBN: 0-387-94823-6. DOI: 10.1007/978-1-4419-8566-8. URL: http://dx.doi.org/10.1007/978-1-4419-8566-8.

Klein, Felix. (1956). *Lectures on the Icosahedron and the Solution of Equations of the Fifth Degree,* Revised. Translated into English by George Gavin Morrice. New York, NY: Dover Publications, Inc., pp. xvi+289.

Miranda, Rick. (1995). *Algebraic Curves and Riemann Surfaces.* Vol. 5. Graduate Studies in Mathematics. Providence, RI: American Mathematical Society, pp. xxii+390. ISBN: 0-8218-0268-2.

Munkres, James R. (1975). *Topology: A First, Course.* Englewood Cliffs, NJ: Prentice-Hall, Inc., pp. xvi+413.

Silverman, Joseph H. and John Tate. (1992). *Rational Points on Elliptic Curves.* Undergraduate Texts in Mathematics. New York: Springer-Verlag, pp. x+281. ISBN: 0-387-97825-9. DOI: 10.1007/978-1-4757-4252-7. URL: http://dx.doi.org/10.1007/978-1-4757-4252-7.

Wilf, Herbert S. (2006). *Generatingfunctionology.* 3rd ed. Wellesley, MA: A. K. Peters, Ltd., pp. x+245. ISBN: 978-1-56881-279-3; 1-56881-279-5.

Index